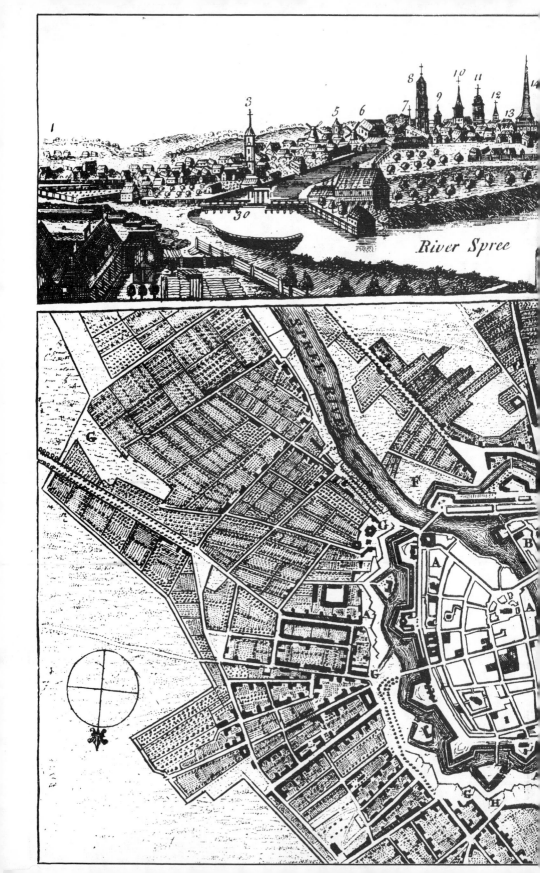

River Spree

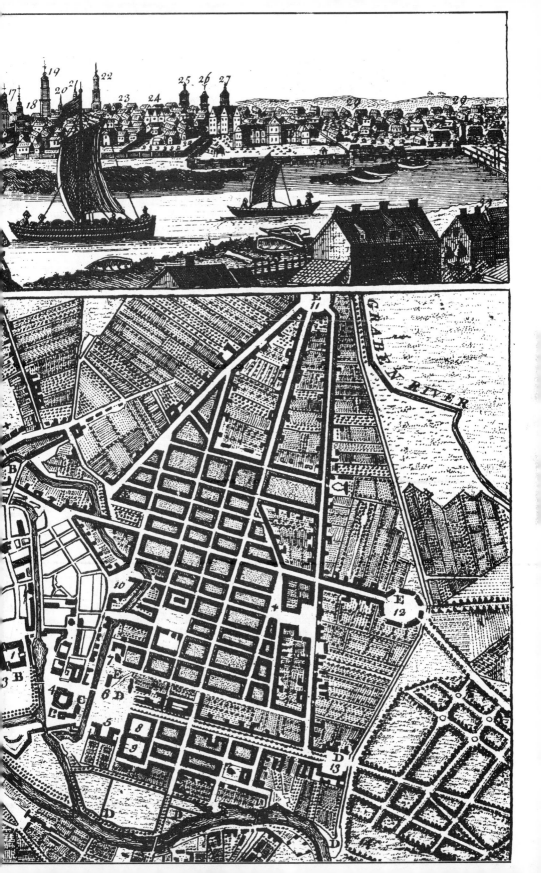

Frontispiece: Plan and prospect of the City of Berlin, *c.*1737. Reproduced from the *London Magazine* 29 (November) 1760, by permission of the Bodleian Library, Oxford (HOPE ADDS 415).

The plan was published on the occasion of the occupation of Berlin by the combined Austrian and Russian forces in 1760, during the Seven Years' War, but like so many others of its kind is obviously based on G. Dusableau's revised Berlin plan of 1737. It will be apparent that, contrary to usual practice today, the plan has north at the bottom of the page, and thus appears 'upside-down'. It shows the elaborate fortifications erected *c.*1685 by the Great Elector enclosing the medieval towns of Berlin (left), Cölln (centre) and the small extension of Friedrichswerder (right). To the west (right) are the planned 'new towns', Dorotheenstadt and Friedrichstadt, while particularly to the north (bottom) and east the irregular growth of suburbs outside the walls is clearly shown.

The eighteenth-century 'Explanation of the plan' is repeated below, with some comments:

A Berlin city; **B** Cullen city (Cölln); **C** Frederic's island (Friedrichswerder); **D** Dorothy stadt or new city (Dorotheenstadt); **E** Frederic's stadt or new city (Friedrichstadt); **F** Cullen (Cölln) suburb **G** Berlin suburb; **H** Spandau suburb; **1** The royal castle (Stadtschloss, Berlin Palace); **2** The royal stables (Marstall); **3** The parade (Lustgarten); **4** The arsenal (Zeughaus); **5** Prince Henry's palace (from 1810, the University); **6** The opera house (Deutsche Staatsoper); **7** The new popish church (St Hedwig's Catholic Cathedral); **8** The great stables; **9** The observatory and anatomy hall (with the great stables, now replaced by the Academy of Sciences and German State Library; **10** The Jerusalem place (Hausvogteiplatz); **11** The circle (Belle-Alliance-Platz, now Mehringplatz, at the Halle Gate [Hallesche Tor]); **12** The octagon (Leipziger Platz, at the Potsdam Gate); **13** The square (Pariser Platz, at the Brandenburg Gate).

The 'Explanation of the view' included the following notable buildings:

1 The royal summer palace of Schonhaus (Schloss Niederschönhausen); **3** The church and the new steeple in Spandau suburb (Sophienkirche); **4** The royal pleasure house and garden of Mon Bijoux (the former Schloss Monbijou was not rebuilt after wartime damage); **5** St George's church (Georgenkirche on Alexanderplatz, demolished in 1950 after wartime damage); **7** Holy Ghost hospital and church (Heiliggeistkapelle, now part of Economics Faculty, Humboldt University); **8** St Mary's Church (Marienkirche); **9** Grey Fryar church (Klosterkirche, of which landscaped ruins remain); **10** Reformed parochial church (Parochialkirche, restored after war damage); **12** Berlin stadt house (predecessor of present 1861–70 Rathaus); **14** St Nicholas church (Nikolaikirche; the oldest church of Berlin, restored after war damage 1987; it received its present second spire only in 1880); **15** The royal castle (Stadtschloss, Berlin Palace); **16** The arsenal (Zeughaus); **17** The royal castle and dome church (Dom, Protestant Cathedral); **19** St Peter's church (Petrikirche, oldest church of Cölln, demolished in 1963 after wartime damage); **23** Frederic's stadt new church (Deutsche Dom); **24** Frederic's stadt French church (Französische Dom); (The prominent towers of the above two churches were added in 1736–82, and do not appear in the panorama); **26** Dorothy stadt church (demolished in 1965 after wartime damage); **29** The new buildings of Dorothy and Frederichstadt; **30** The Wiedendam bridge.

The panorama appears to have been taken from a point on the Spree downstream of the city near to the letters 'D' at the bottom (north) of the plan. Note the depiction of shipping on the river and boat-building along the banks (hence the street-name 'Schiffbauerdamm', which became well known through the occupation by Berthold Brecht's *Berliner Ensemble* of the 'Theater am Schiffbauerdamm').

Berlin

BER|LIN

The spatial
structure of
a divided city

T. H. ELKINS

with B. Hofmeister

METHUEN | London and New York

First published in 1988 by
Methuen & Co. Ltd
11 New Fetter Lane, London EC4P 4EE

Published in the USA by
Methuen & Co.
in association with Methuen, Inc.
29 West 35th Street, New York NY 10001

Set by Hope Services, Abingdon
Printed in Great Britain at
the University Press, Cambridge

British Library Cataloguing in Publication Data
Elkins, T.H.
   Berlin: the spatial structure of a divided city
   1. Berlin (Germany)—History
   I. Title   II. Hofmeister, B.
   943.1'55   DD860
ISBN 0–416–92220–1

Library of Congress Cataloging in Publication Data
Elkins, T. H. (Thomas Henry)
   Berlin: the spatial structure of a divided city /
   T.H. Elkins with B. Hofmeister.
      p.   cm.
   Bibliography: p.
   Includes index.
   ISBN 0–416–92220–1 (cased)
   1. Berlin (Germany)—History.   I. Hofmeister, Burkhard.
   II. Title.
   DD866.E44 1988
   943.1'55—dc19

Contents

Plates

Figures

Tables

Preface

Anyone attempting to write about both sides of Berlin enters a political minefield; at any moment some apparently neutral statement of fact may suddenly be challenged as having political undertones, and even such seemingly innocuous expressions as the names of places are revealed as value-laden. This being so, it is essential to make clear where responsibility for the text that is to follow lies. Without the assistance of Professor B. Hofmeister of the Technical University, Berlin, this book would never have been written. He has allowed his own extensive writings on West Berlin to be drawn on, he has suggested numerous sources of information, and has read through the text, making suggestions for improvement, removing errors of fact, and correcting German orthography.

Without this support, the author would not have dared to begin work, yet it has to be emphasized that the whole concept of the book, its writing, and all value judgements contained in it are his responsibility and his alone. He carries sole responsibility for whatever errors remain in the text, and in particular is solely responsible for the sections of the text dealing with East Berlin.

The selection of appropriate forms of place names involves difficulties, partly because they must be intelligible to English-language readers, partly because, where alternative versions exist, they may be politically 'loaded'. The term 'Germany' is taken in a general sense of indicating the area occupied by German-speaking people until the nineteenth century, but from 1871 as meaning the territory included at any one time within the pre-1939 boundaries of the German Reich. The internationally accepted designations 'Federal Republic of Germany' (abbrev. 'Federal Republic') and 'German Democratic Republic' (abbrev. 'GDR') are employed. Where there is an accepted English form for states, Prussian provinces, other large areas and physical features spanning political boundaries, this will be used; to write of 'Preussen' rather than the familiar 'Prussia' would be unacceptably pedantic. As for the names of towns, it was the initial intention to follow internationally agreed usage and to employ the form current in the state concerned. As writing proceeded, it became apparent that while the correct forms for German towns were reasonably acceptable (nobody is going to be confused by 'Hannover' or 'München'), the policy would involve the use of forms for cities outside Germany that are not used in everyday English speech: 'Torino' or 'Genève' seem unacceptably precious. So there has been a fall-back on expediency; where customary English-language names exist, they are used. It should however be noted that in the historical section of the book German forms are used for towns in areas once occupied by a German-speaking population which now lie outside German control and bear other names.

Divided Berlin presents particular problems. In the view of the Federal Republic, the accepted nomenclature is Berlin (West) and Berlin (East). Even this apparently innocent, purely geographical terminology is in fact value-laden, as it may be held to support the position, common to the Federal Republic and to the three western occupying powers (United Kingdom, France, and the United States), that the city is a unity under a special regime, and just happens to be divided into western and eastern portions. This is a viewpoint countered at every possible opportunity by the GDR, which insists that the eastern part of the city is an integral part of its territory and indeed its capital, and so officially bears the title 'Berlin Capital of the German Democratic Republic' (Berlin Hauptstadt der Deutschen Demokratischen Republik). The author is clearly placed in a difficult position; whatever terminology is chosen will be liable to criticism. A weak resort to expediency seems the only way out. The official GDR title is both extremely cumbersome and totally unfamiliar to the vast majority of English-language readers, who daily in their newspapers read of 'East Berlin', just as they read of 'West Berlin'. These names will accordingly be used, and it is no doubt futile to insist that they are intended only as convenient labels expressing geographical location, not as some kind of political proclamation.

Acknowledgements

The author wishes to acknowledge his debt to the researches and writings relating to Berlin of geographers and others far too numerous to mention individually, that have been drawn on in the preparation of this book. Each of its chapters could have been the subject of a monograph or book, and frequently has been; in the space of a single volume, it is only possible to attempt an overview, which to the various specialists must appear superficial.

The author wishes in addition to acknowledge assistance for study visits to Berlin received, at various times, from the German Academic Exchange Service (DAAD), the Economic and Social Research Council, the Academy of Sciences of the German Democratic Republic (on the nomination of the British Academy) and the GDR Ministry for Universities and the Humboldt University (on the nomination of the British Council). Even when such visits were devoted to other research objectives, they contributed to a build-up of familiarity with the city and its problems. The author also wishes to thank all those members of the Free University, the Technical University, the Humboldt University, the GDR Bauakademie and the Hochschule für Ökonomie in Berlin-Karlshorst who have at various times given of their time in interviews and on excursions to demonstrate aspects of the development and nature of the city.

For permission to reprint material, acknowledgement is made to the following: to K Baedeker Verlag (Freiburg) for extracts from their 1936 English-language *Germany*; to the *Geographical Magazine* for the reproduction of material on the Berlin S-Bahn; to B. Hofmeister and Wissenschaftliche Buchgesellschaft, Darmstadt, for figures 6.3, 7.2, and 7.3; to B. Hofmeister and D. Reimer Verlag for figures 4.2 and 6.5; to B. Hofmeister and Colloqium Verlag, Berlin for permission to use a modified version of a map from *Von der Residenz zur City; 275 Jahre Charlottenburg* (Berlin 1980) as the basis of figure 7.1; to B. Backé and *Berichte zur deutschen Landeskunde* for figure 8.2; to

H. Müller and G. Westermann Verlag for figure 8.6; to G. Wettig and Westview Press for permission to draw on his contribution to R. D. Francisco and A. Merritt (eds) *Berlin Between Two Worlds* (Boulder, Colo. and London, 1986); to the Bodleian Library for permission to reprint a *c.*1737 map of Berlin (Hope Adds. 415) as frontispiece.

Susan Rowland (Sussex) drew figures 2.1, 5.3, 6.1, 6.2, 6.4, 7.5, 8.1, 8.2, 8.3, 8.4, 8.5, and 8.6; Angela Newman (Oxford) 1.2, 3.1, 4.1, 4.3, 4.4, 4.5, 5.1, 5.2, 7.3, and 7.4; H.-J. Nitschke & G. Braun (Berlin) 4.2, 6.3, 6.5, 7.1, and 7.2. The photographs are by the author.

Abbreviations

ARWOBAU	Arbeitnehmer-Wohnheimbaugesellschaft mbH (West Berlin: provides accommodation for immigrant workers from the Federal Republic)
AVUS	Automobil-Verkehrs-und-Übungsstrasse (prototype autobahn)
BEWAG	Berliner Kraft- und Licht AG (electricity supply company)
BIG	Berliner Innovations- und Gründerzentrum (West Berlin Centre for Industrial Innovation and Promotion)
BRD	Bundesrepublik Deutschland (see GFR)
BVG	Berliner Verkehrsgesellschaft (unified public-transport organization)
CDU	Christlich-Demokratische Union (Christian Democratic Union)
DDR	Deutsche Demokratische Republik (German Democratic Republic)
DM	Deutsche Mark (German Mark, in Federal Republic)
EEC	European Economic Community
EWG	Europäische Wirtschaftsgemeinschaft (European Economic Community); see EEC
DB	Deutsche Bundesbahn (German Federal Railways)
DR	Deutsche Reichsbahn (German State Railways, in GDR)
FDGB	Freier Deutscher Gewerkschaftsbund (Free Federation of German Trade Unions, in GDR)
FDJ	Freie Deutsche Jugend (Free German Youth, in GDR)
FDP	Freie Demokratische Partei (Liberal Democratic Party)
Federal Republic	German Federal Republic (Bundesrepublik Deutschland)

FU	Freie Universität (Free University)
GASAG	Berliner Gaswerke (gas supply company)
GDR	German Democratic Republic (Deutsche Demokratische Republik)
GEHAG	Gemeinnützige Heimstätten Spar- und Bau A.G. (a housing association or building society)
GFR	German Federal Republic (form found in GDR English-language publications, and sometimes elsewhere: 'Federal Republic' is here preferred)
IBA	Internationale Bauaustellung (International Building Exhibition)
ICC	Internationales Congress Centrum (International Congress Centre, in West Berlin)
KPD	Kommunistische Partei Deutschlands (German Communist Party; in the Federal Republic only; was dissolved, and succeeded by DKP – Deutsche Kommunistische Partei)
LSG	Landschaftsschutzgebiet (protected area of outstanding natural beauty)
M	Mark (German Mark, in GDR)
NEG	Naherholungsgebiet (short-period recreational area in GDR)
NSG	Naturschutzgebiet (nature reserve)
RIAS	Rundfunk im Amerikanischen Sektor (Radio in the American Sector)
S-Bahn	Stadtbahn (city rail system)
SED	Sozialistische Einheitspartei Deutschlands (United Socialist Party of Germany, in GDR)
SEW	Sozialistische Einheitspartei Westberlins (West Berlin equivalent of SED)
SFB	Sender Freies Berlin (broadcasting station of Free [West] Berlin)
SMA, SMAD	Soviet Military Administration
SPD	Sozialdemokratische Partei Deutschlands (Social Democratic Party of Germany, in Federal Republic and West Berlin)
SSD, STASI	Staatssicherheitsdienst (State Security Police, in GDR)
TSI	Treuhandstelle für Industrie und Handel (Office of Trustee for Industry and Trade, in West Berlin)
TU	Technische Universität (Technical University)
U-Bahn	Untergrundbahn (underground railway, subway; sections are at surface or even elevated)
VOPO	Volkspolizei (People's Police, in GDR)

Introduction

The existence of Cölln, one of the twin towns which formed the original nucleus of the present great city of Berlin, is first recorded from 1237, a date which is conventionally taken as that of the foundation of the city as a whole. The 1987 celebrations on the 750th anniversary of this date were the occasion for the publication of the present book.

It is an inevitable consequence of the present political situation in Berlin that any such celebration must take place separately in the two halves of the city. Berlin, while having many aspects of its life in common with other great cities throughout the world, is an enormous geographical exception, a divided city. What is more, while East Berlin has relatively normal access to its hinterland in the German Democratic Republic, West Berlin is to a considerable degree sealed off both from East Berlin and from its natural hinterland; its links are essentially with the Federal Republic, 175 km or more distant. The evolution of this extraordinary situation will be followed in greater detail below (chs 1–2), but the implications of division and insularity run through every section of the book.

Berlin: product and victim of history | 1

1.1 The rise to capital status

1.1.1 No outstanding natural advantages

At the heart of modern Berlin, the dark, polluted waters of the river Spree slide malodorously round the island of Cölln, that once contained the Berlin Palace of the rulers of Prussia and of the German Second Reich, now replaced by the glass-fronted modernity of the Palast der Republik of the GDR, a building that houses the country's parliament (see figs 6.1 and 6.2). Here, where the stream once divided into three branches and was further interrupted by sandbanks, was from the earliest times the lowest easy crossing of the Spree before it emptied into the wider and often lake-like Havel river. It lay at a point where the low glacial-drift plateaus of Barnim to the north and of Teltow to the south, zones of relatively easy movement by land, were less than 5 km apart across the course of the Berlin *Urstromtal* (former pro-glacial melt-water channel) within which the river flows. In this part of its course the Spree is clear of the lakes which clutter the Havel confluence in the neighbourhood of Spandau to the west and the Dahme confluence in the neighbourhood of Köpenick to the east, lakes which were later to become a precious recreational resource for the future world city (see fig. 3.1). It must not be forgotten that the Spree, as well as being an obstacle to be overcome, was also a means of movement by water, as it remains to this day. When frozen in winter it could also be used for movement by sledge (Cornish 1923: 153–5).

If the geographical situation of Berlin had some advantages, its site was not particularly suitable for building purposes. The sands which form the floor of the *Urstromtal* are intermixed with peat, forming a particularly treacherous foundation. The earliest nuclei, from which medieval Berlin coalesced, clung to sandbanks, still marked today by the city's surviving medieval churches, such as the Marienkirche, which now shares its 'island' with the East Berlin TV tower. Problems with the foundations of buildings have continued throughout the city's history.

Cölln, on its island, was one of the numerous towns founded in the course of the German eastern colonization. The year 1237, when it was first recorded, is conventionally accepted as the date of the foundation of the whole city. Its twin town of Berlin, on the north bank, was first recorded in 1244, but archaeological evidence suggests a Germanic occupation of both Cölln and Berlin at least eighty years earlier than indicated in documentary sources, perhaps in the decade 1160–70. Unlike some other German town sites in the region, Berlin-Cölln does not appear to have been preceded during the Slav period by any significant settlement.

The advantages of the Spree crossing were not overwhelming, even in the medieval period. Within the area of what was to become Greater Berlin there were two earlier sites of urban or proto-urban nature. To the west, Spandau was a fortified Slav settlement dating from the end of the eighth or the beginning of the ninth century AD, commanding the Spree–Havel confluence. Similarly, the Slav settlement of Köpenick to the east commanded the Dahme–Spree confluence (see section 6.4.4.). Between the two, Berlin had no particular prominence among the German towns that were established between the Elbe and Oder rivers, and lay away from the major medieval trade routes. What can perhaps be said is that as soon as the plateaus of Teltow and Barnim became the scene of organized German village settlement, their easy accessibility enabled Berlin-Cölln to organize a coherent local market area, whereas Spandau and Köpenick were hemmed in by the lakes which, at an earlier stage, had been a defensive advantage. It was also possible at Berlin to dam one of the branches of the river to power a public mill, still recorded in the street name Mühlendamm. Nevertheless, towns a little further afield, such as Magdeburg or Frankfurt on Oder, were certainly better endowed by nature than Berlin-Cölln. We must look to other causes than the natural environment for a sufficient explanation of the rise of the city to world status.

1.1.2 Berlin's geographical situation

The advantage to a capital of having a central situation within the territory of a state can, perhaps, be overstressed; it is not difficult to compile a list of capitals that have apparently managed to function quite adequately from markedly asymmetrical positions. Certainly, Berlin has rarely been centrally placed with regard to the fluctuating expanses of German territory that the city has from time to time controlled. Even within the Germany of the Second Reich of 1871,

Berlin was not far off 100 km closer as the crow flies to its furthest north-eastern outpost at Memel (now Klaipeda in the Soviet Union) on the Baltic than to its south-western extremity in Alsace, although on the other diagonal the cities of industrial Upper Silesia and the Danish frontier were approximately equidistant. The biggest disparity was on Berlin's own latitude, where the boundary with Russian Poland to the east was about 280 km distant, whereas the Netherlands frontier in the other direction was over 400 km away (fig. 1.1). The asymmetrical location of Berlin has been accentuated by the results of two world wars: whereas the Netherlands boundary has scarcely changed, the boundary with the territory occupied by Poland as a result of the Second World War lies less than 80 km distant in the direct line. By autobahn to the official crossing point it is about 100 km, a little further, but still only about an hour and a half's journey even at the sober pace of GDR road traffic (fig. 1.2). Within the reduced area of the German Democratic Republic, Berlin is much more centrally placed

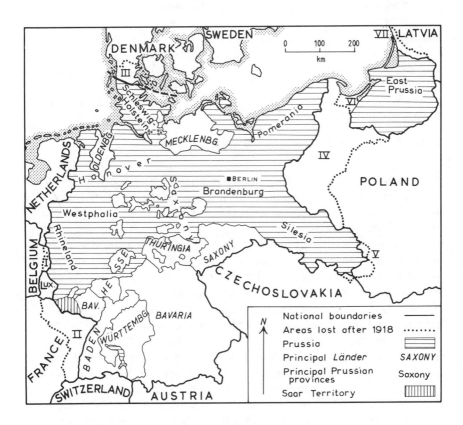

Figure 1.1 Political divisions after the First World War

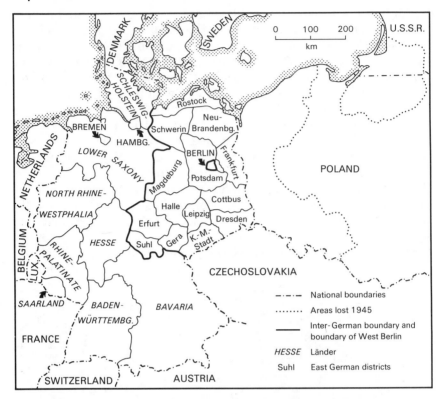

Figure 1.2 Political divisions after the Second World War

than within the pre-1914 Reich; nevertheless, the southern crossing into Bavaria is some 275 autobahn km away (at present five hours or more by train) as compared with the proximity of the boundary with Polish territory to the east.

It is more meaningful to relate the geographical situation of Berlin not to distance in an absolute sense but to the distribution of population, which can broadly be taken as an indicator of the distribution of economic activity and economic opportunity. Central Europe has two major axes of population and urban density. One follows the Rhine from Switzerland to its mouth, the second intersects the first in the Ruhr area and runs somewhat south of east through Hanover to Saxony and Silesia (fig. 1.3). It might be expected that an ideal German capital would be located somewhere on these axes, preferably at their intersection, whereas Berlin is situated in their north-east quadrant, in the relative emptiness of the North German Lowland. Within this general area it has not even the advantage of a situation on the Baltic coast. What can perhaps be said is that in the narrower context of the GDR the location of Berlin on the Northern Lowland provides a valuable spatial counterweight to the cities of industrial Saxony, contributing towards meeting the country's spatial-planning goal of balanced regional development.

Berlin's relative isolation from other major centres of population and

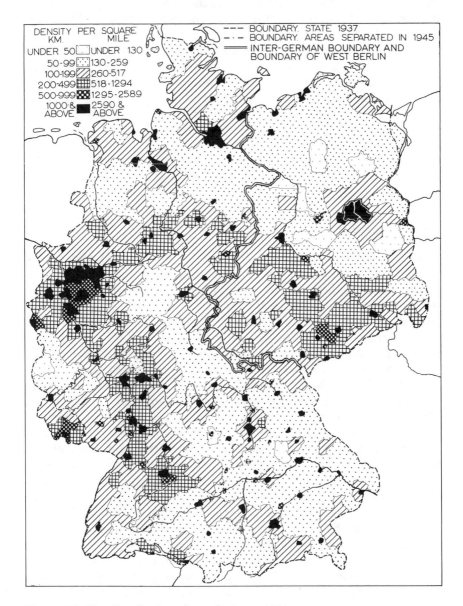

Figure 1.3 The distribution of population, c.1950.

economic activity is however mitigated by the lack of natural obstacles in the terrain that surrounds it; once the city began to develop in importance, it was able to establish the necessary external linkages without insuperable difficulty. Although Berlin was a member of the Hanse at the latest by 1359 (Vogel 1966), it was not on any of the really major routes running southwards from the coast, and it lay north of the great east–west routes along the Harz Foreland and through Thuringia. Yet once Berlin began to assume importance as a residence of the rulers of Brandenburg, there was no difficulty in diverting the post roads through the city and the adjoining palace town of Potsdam. Similarly, there was no obstacle to making Berlin the heart of the nineteenth-century Prussian rail system, and the point of origin of the national autobahn system in the 1930s. Only with waterways has nature played a more positive role; the combination of shallow, generally east–west trending *Urstromtäler*, north–south breakthrough stretches (such as those occupied by the Havel) and a profusion of lakes facilitated a steady improvement of river navigation and the building of canals from the eighteenth century onwards.

1.1.3 The product of history

It is impossible to dissociate the rise of Berlin from the rise of Brandenburg-Prussia to hegemony in Germany. Initially, the dual settlement of Berlin-Cölln was only one component in the totality of colonization measures initiated in what became the Mark of Brandenburg, designed to increase the power, the lands, and the tax revenues of the ruling Askanier family. Although the Askanier retained a residence in Berlin, the town was not one of the bastions of their power, unlike the fortress at Spandau. Berlin-Cölln was essentially a central market settlement for the planned villages of Barnim and Teltow, into which were concentrated both colonizing Germans and the indigenous Slavs, who had formerly lived in small, scattered settlements. Some of the earliest German villages appear to have originated at the turn of the twelfth and thirteenth centuries under the protection of the Knights Templar (hence the place-name Tempelhof), whose patroness was the Virgin; their villages Mariendorf and Marienfelde (among others) are part of Greater Berlin today. A mint had also been established by 1280, the most important in the Mark of Brandenburg.

The lords of the Mark were also prepared to confer a range of taxation and trading privileges, mostly, no doubt, in return for hard cash. For example, merchants passing through the town were obliged to offer their goods for sale or pay disproportionately high customs dues. It also appears that Berlin merchants were not obliged to pay customs dues to the Margrave on goods moving within the Mark of

Brandenburg (Vogel 1966). The citizens grew wealthy, trading rye and timber with the ports of the North Sea coast, although social contrasts opened up between a ruling group of rich merchants and the lesser traders and craftsmen, organized in guilds.

The death of the last Askanier Margrave in 1319 ushered in a period when the Mark of Brandenburg passed through the hands of a variety of frequently absent rulers; this was also a period when Germany as a whole was in a period of political disintegration. Like other towns throughout Germany, Berlin was able to use this period of confusion to bid for increased influence and self-government. Considerable land holdings were built up in the surrounding villages, and further taxation and legal privileges were bought from the various rulers. Berlin and Cölln increased their links, especially for defensive purposes, building a third town hall for common purposes on the bridge that linked them together. Berlin became a leading member of a league of towns of the central Mark, as well as a member of the Hanse.

Independence was, however, not to last, yet ironically its loss was to ensure that Berlin would rise to be a great capital rather than fall into the political impotence of former great medieval free cities such as Augsburg. In 1411 the emperor appointed Burgrave Frederick VI of Nuremberg, a member of the House of Hohenzollern, to be governor of the Mark of Brandenburg; in 1415 he was definitively installed as Elector Frederick I.

The burghers of Berlin attempted to maintain their independence, looking for help to the Hanse, to the link with Cölln and to association with neighbouring towns in the Mark, but it was soon clear that a new era of direct princely rule had begun. The next elector, Frederick II (Irontooth), systematically eroded the privileges of the town, in part by playing on internal dissensions between the richer merchants and the rest of the population. The elector gained the right to receive the keys of the town on demand and to approve the burgomasters and members of the two town councils. Entry into leagues with other towns was forbidden, and the links with Cölln again dissolved. The commercial, customs, and legal privileges that had been granted to the town over the years were taken back into the increasingly absolutist hands of the ruler, who installed his judges in the former common town hall on the bridge linking the dual towns.

1.2 Electoral and royal capital

The new order was symbolized by the building between 1443 and 1451 of the *Stadtschloss*, the electoral town palace, on the island of Cölln. From about 1470–80 this was continuously the seat of the electoral court, Berlin-Cölln thus entering on a new career as a *Residenzstadt*. Although clearly supreme in Brandenburg, of which 'it was now administratively an integral part, it was at this stage only one among many petty princely capitals in a politically fragmented Germany. That it was to rise above all these others was the result of the linkage of its fortunes with those of the House of Hohenzollern.

The transformation of Berlin-Cölln into a princely capital brought about a parallel transformation in the economic, social, and cultural life of the dual town. It came to be dominated by officials, who were under the legal jurisdiction of the court, not of the town, and who were exempt from the taxes and other obligations of citizenship. Initially, the Landtag (the parliament) met in the palace, which also housed the electoral Chancellery, the higher courts of justice, and the administrations for taxation, finance, and ecclesiastical affairs (Brandenburg accepted the Reformation in 1529–40, a move giving religious, economic, and political advantages to the ruler). What the new capital did not receive was Brandenburg's university, which was instead founded in 1506 in the trading town of Frankfurt on Oder. Readers in England will be well aware that this was not the only example of university location outside the capital; perhaps the Electors were fearful of student unrest, or desired to keep an uncontrollable foreign element at a safe distance (Schinz 1964: 66). The demand for goods to supply the court, its officials, and the army stimulated the growth of manufacturing. The richer merchants increasingly turned from long-distance trade to the more immediately rewarding business of meeting electoral needs for goods and credit. The number of inhabitants rose from 6,000 in 1450 to 12,000 at the end of the sixteenth century, in spite of periodic visitations of the plague.

By this time, the initial palace complex had been completed. An elaborate Renaissance wing had been added, overlooking a parade ground, and linked by a wooden gallery to the *Hofkirche*, the electoral church. Other necessities such as a guardhouse, stables and a prison were also provided. The situation of this complex in formerly neglected, swampy land at the north-west fringe of Cölln was to be of fundamental importance for the development of Berlin in centuries to come, determining that the western sector would be occupied by the 'polite' inhabitants of the city, and by the services that they required.

The subsequent history of Brandenburg was far from peaceful; there were severe reverses to come, but somehow (until the final fateful year of 1918) the Hohenzollerns always managed to survive each crisis and, usually, to emerge with increased territorial possessions. A particularly low point came in the first half of the seventeenth century. Admittedly, the acquisition in 1614 of Kleve (Cleves) and the Mark in the Rhinelands were potentially significant in terms of laying the foundations of the subsequent Rhine Province, and in 1618 Johann Sigismund united the Duchy of Prussia to the Electorate, but these events were overshadowed by the outbreak of the Thirty Years' War. Berlin suffered less than the rural areas from the direct impact of military operations; unlike Magdeburg it was spared the horrors of being taken by storm, but instead had to endure occupation by a succession of armies. The inhabitants suffered by having troops billeted on them, by having their horses and cattle commandeered, and by having to contribute to repeated subsidies. Loss of trade, inflation, scarcity, and disease added to their problems. Late in the war, houses in the suburbs that had grown up outside the old walls were destroyed to prepare for a siege which fortunately never came. When the young Elector Frederick-William (1640–88) returned to Berlin, it was to a ruined city that had lost half its population, but in common with the other German princes he gained from the war the abolition of the vestigial control of the Empire over his absolute power.

Frederick-William is regarded as having laid the foundations of Brandenburg-Prussia, that improbable state that emerged not from geography, not from the logic of history, not from common ethnicity or the common interests of its inhabitants but which was 'the creation of a few kings possessed by the fury of *raison d'état* and of the servants whom they commanded' (Mann 1974: 32–3). In the period 1658–83 Frederick-William caused the city to be surrounded by a belt of fortifications better reflecting the needs of the age of artillery than the old medieval wall (see fig. 6.2). These fortifications appear to have discouraged attack by the Swedes, allies of Louis XIV in his war of 1672–78 in the Low Countries, who invaded Brandenburg but failed in an attempt to take Spandau by storm. They were eventually defeated by Frederick-William at Fehrbellin, north-west of Berlin, in 1675, after which victory he was known as the Great Elector (his equestrian statue by Andreas Schlüter is a triumph of baroque sculpture and one of the artistic treasures of the city; it formerly stood on the Lange Brücke in front of the now destroyed Berlin Palace and now stands before the Charlottenburg Palace).

The permanent presence of a standing army caused inconvenience to the citizens, who had either to have soldiers billeted on them or pay for exemption, but military requirements stimulated the growth of manufacturing and a rapid rise of population, which reached 61,000 by 1710. Another contribution both to the economy and to the composition of the Berlin population was provided by immigration, notably of Calvinist religious refugees (Huguenots) from the France of Louis XIV. By the end of the sixteenth century every fifth Berliner was of French extraction, forming a largely self-governing community, with its own churches and schools. The French Church (Französischer Dom) was established on the Platz der Akademie (the former Gendarmenmarkt, earlier Friedrichsstädter Markt). Painstakingly restored from severe wartime damage, it stands as a witness to the importance of the immigrants. The Huguenots introduced numerous new branches of manufacture and in addition to their impact on the economy, strongly influenced administration, the army, the advancement of science, education, and fashion. The Berlin dialect still employs many terms of French derivation. The increase in the number of officers and court officials, together with the general increase of the population, led to the initiation of a process of adding new quarters to the city, notably along the axis of the Unter den Linden to the west (see section 6.2.2). Administratively, these took the form of independent 'new towns', laid out according to the plans of the ruler, on land for the most part owned by him (see figs 6.1 and 6.2).

The succeeding elector, Frederick III, assumed the royal title of King Frederick I in respect of his Prussian territories, which lay outside the Empire, but the name 'Prussia' (Preussen) was subsequently extended to the whole of Brandenburg territory, including its Rhenish portions. Berlin thus became the capital of a monarchy, a role it was to maintain until the close of the First World War. Urban expansion continued; the Berlin Palace was rebuilt in baroque style, and many other public buildings and churches were created. The Academy of Arts was founded in 1696, and in the following year the philosopher Leibniz initiated the Academy of Sciences. A particularly brilliant artistic and intellectual circle gathered around Frederick's consort, Sophie Charlotte, at her new palace at Lietzenburg west of the city, which after her death in 1705 was renamed Charlottenburg. In 1709 Berlin, Cölln, and the three 'new towns' on the Unter den Linden axis were administratively united to form a united capital city of Berlin. It was to be of enormous importance for the future development of the admittedly still small (1,343 hectares) city that as much as 45 per cent (1786) of its area lay in royal ownership, its planning directly under royal control (Hofmeister 1985).

Under Frederick-William I, the 'soldier king', the palace pleasure grounds, the Lustgarten, were turned into a parade ground, and in 1735 the whole city was surrounded by an outer pallisade or wall. This was not designed for defensive purposes, a function allocated to Spandau, but to prevent the flight of deserters and to facilitate the collection of the octroi, the customs duty on goods entering the city. Enough open land was included to meet the building needs of the city well into the nineteenth century. The gates in the new wall are recalled to this day by place-names ringing the city's historic core. The gates on the 'polite' western side of the city were provided with formal squares or circuses: the Rondell (later Belle-Alliance-Platz, later

Mehringplatz) at the Halle Gate, the Octogon (later Leipziger Platz) at the Potsdam Gate, and the Quarré (later Pariser Platz) at the Brandenburg Gate, the best known of them all.

The reign of the philosopher-soldier Frederick II (Frederick the Great), which lasted from 1740 to 1786, at first brought tribulation. Taxes had to be raised to finance his expansionist ambitions; in the darkest days of the Seven Years' War the city was briefly occupied by the Austrians in 1757, and in 1760 bombarded and then occupied by the Austrians and Russians. On both occasions a heavy subsidy had to be paid by the citizens. Yet by luck and good management Frederick managed to emerge from the war with his gains in Silesia and Poland intact and his country undisputedly a major actor on the European stage. Retrospectively, he may appear as an absolutist despot, but at the time Prussia was looked upon as progressive and enlightened:

> It was thought to be well ruled because it was ruled in the interest of the state by a strict professional king and a reliable, professional bureaucracy. Its administration was rational, logical and honest, its justice prompt and impartial, approaching what was later called the reign of law. (Mann 1974: 34)

The freedom that Frederick permitted in religion and philosophical speculation was not, however, extended to the political sphere.

The advancement of the state and its enhanced administrative efficiency were reflected in its capital. Although Frederick preferred to live in his Potsdam palaces, he did not neglect Berlin. With royal encouragement manufactures blossomed, the Academies of Arts and of Sciences were re-founded and, particularly in the peaceful second half of the reign, the city was ornamented with new public buildings, including a new cathedral on the Lustgarten. A start was made with the building of proper barracks for the substantial garrison, replacing the unpopular system of quartering. The city administration was reformed, although it lay strictly under the control of the royal officials. The new director of police imposed a degree of public order and safety that was the envy of other European cities. By the time of Frederick's death in 1786 the population of the city had risen to about 150,000, similar to Amsterdam, Madrid, Rome, and Vienna, and substantially exceeded in Europe only by London and Paris (Vogel 1966: 102). The nineteenth-century equestrian statue of Frederick the Great, banished on political grounds in 1950 to the recesses of the Sanssouci Park at Potsdam, was in 1980 restored by the GDR government to its place of honour on the Unter den Linden, but some commentators still regard Frederick's almost magical legend as the great man who could get away with anything, as having provided an unfortunate precedent for the German people.

Plate 1.1 The statue of Frederick the Great, restored to its original position on the Unter den Linden, East Berlin.

As the peaceful and prosperous city coasted gently towards the events of 1789, artistic and intellectual life flourished in an atmosphere of enlightenment and gentle romanticism. The revolutionary ideas which left such an indelible mark on German thought and German history were at first received sympathetically in Berlin. In 1795 Prussia abandoned opposition to revolutionary France and retreated into eleven years of neutrality in the midst of conflict. It was a period when the capital witnessed a continuation, even enhancement, of the explosion of intellectual and artistic creativity that had marked the later eighteenth century, exemplified in the triumphant productions of Schiller's plays in the Schauspielhaus on the Gendarmenmarkt square. Unfortunately, Prussia's renewed entry into the struggle against Napoleon was ill-timed, ending in crushing defeat at Jena in 1806. Napoleon installed himself in the Berlin Palace, the Quadriga from the Brandenburg Gate was removed to Paris as a trophy and the city settled down to a gruelling two-year occupation. By the time that the French left after imposing on Prussia the humiliating Peace of Tilsit in 1808, the city was economically shattered and burdened with a debt that was not paid off until 1861 (Vogel 1966: 109–10).

The story of the revival of Prussia and the internal reforms carried through under the inspiration of Freiherr vom Stein is well known. His most enduring achievement was urban reform, notably the introduction of local self-government by elected councils. Berlin was given a city council (*Stadtverordnetenversammlung*) elected on a much broader base than formerly; its executive organ was the *Magistrat*. Local democracy was not, however, total; the election of the *Bürgermeister* and *Oberbürgermeister* was subject to official approval, while police and justice were also kept firmly in the control of the state. It is impossible not to be struck by the contrast between a Prussian city such as Berlin, with town council, professional administrators, building regulations, paved streets, hospitals, and schools, and the English industrial towns rising in the same period, under virtually no control at all.

Another sign of a new spirit was the foundation in 1810 of Wilhelm von Humboldt's Berlin University, the university in Frankfurt on Oder being transferred in the following year to the more distant Breslau. Illuminated by scholars of the eminence of Wilhelm von Humboldt, Ranke, Fichte, Hegel, Schelling, Ritter, or the brothers Grimm, the university not only became the intellectual centre of Berlin life, but achieved the highest of reputations in the German-speaking lands and beyond. The Friedrich-Wilhelm University was installed in the Prinz Heinrich Palace in the Unter den Linden. In 1949 it took the name Humboldt University; in front of the building are statues of Wilhelm von Humboldt and his brother Alexander, the great geographer.

Whereas the states of southern and western Germany rather passively accepted the restructuring and reforms imposed upon them by the French, reform in Prussia took more the nature of an anti-French resistance movement, and leading citizens began to dream of a war of liberation. Berlin had still to endure a further occupation by the *Grande Armée* on its march to Moscow, but yet once more Prussia emerged from disaster on the winning side, participating in the coalition that defeated Napoleon at Leipzig in 1813 and again at Waterloo in 1815. Unlike the states of western and southern Germany, Prussia could be proudly conscious of having saved itself by its own exertions. Admittedly, Prussia had to cede to Russia the Polish territory it had claimed in 1795, but was compensated with economically much more important territories in north-west Saxony, the Rhinelands and Westphalia. Sprawling in this way across northern Germany, Prussia was poised for the final act of forcing unity upon the major part of the German people. This increased significance of Prussia was of course reflected in the increased significance of its capital, Berlin.

In architectural as in political terms, the royal hand still lay firmly on the city; this was the period when K. F. Schinkel gave the area of the Unter den Linden its final form, and P. J. Lenné laid out the Tiergarten. In ruling Prussian circles, the spirit of national revival and fraternity did not outlast the peace; Frederick-William III took up an increasingly reactionary position. He dropped his more liberal advisers, such as Wilhelm von Humboldt, banned some fairly innocuous patriotic associations and, particularly in periods of repression in 1819 and 1830–33, filled Berlin's Hausvogtei state prison with detainees. The arrival on the throne in 1840 of the apparently more liberal but mentally disturbed Frederick-William IV did not exempt Prussia from the impact of the revolution that swept continental Europe in 1848. In a sequence of events that began and remained in confusion, the army attacked revolutionaries in Berlin even after the king had conceded reforms. Frederick-William felt obliged to pay homage to the revolutionary dead in the courtyard of the Berlin Palace; 'only the guillotine is missing now', the queen is said to have remarked. There was also 'the curious procession on horseback which Frederick-William made through the city on 21 March, wearing a sash of black, red, and gold (the revolutionary colours of national Germany), to announce that he was placing himself at the head of Germany' (Mann 1974: 161). Yet once more little came of these reform attempts; reaction returned, and in 1849 the king refused the offer of the German crown at the hands of the German national assembly that had been called in the Frankfurt Paulskirche. German unity was not to come by this democratic route.

As Prussia fastened its control on Germany, so Berlin grew not only in political but in economic predominance. The removal of burdensome customs barriers had begun with the foundation and progressive extension of the *Zollverein*, but Berlin's trading relations with the Baltic ports were still hampered by the Mecklenburg customs boundary, and trading relations with Hamburg by customs boundaries on the Elbe and the inclusion of Lauenburg and Holstein in the Danish customs area. These impediments were swept away in 1866 by the resolution in Prussia's favour of the Schleswig-Holstein question and the formation of the North German Confederation, with direct benefit to Berlin. The city was able to exploit its relatively central situation in a state stretching from Memel to the Rhine, to some extent acting as a trading intermediary between the agricultural east and the economically more developed west.

Improvements made in the 1820s to Prussia's system of main roads (with Berlin as a prominent node) and the efficiency of the long-distance postal service based on horse transport, may explain some initial slowness of railway development. The first German rail line linked Nuremberg and Fürth in 1835, while Saxony had joined its two largest cities, Leipzig and Dresden, by 1839. The first Prussian line of 1838 covered only the short distance between the palace towns of Berlin and Potsdam. Thereafter, the long-distance radials were progressively filled in, keeping to Prussian territory (or at least avoiding Saxon territory) so far as possible: Wittenberg to Köthen (1841), Frankfurt on Oder 1842, and its continuation to Breslau (1846), Stettin (1843), Magdeburg and Hamburg (1846). By 1879 Berlin was the centre of eleven radiating main lines and had become a preponderant rail node within Germany. Internationally, it was a major crossing point, notably of the lines Paris–Warsaw–Moscow and Scandinavia–Vienna–Milan. The railway terminals, those most characteristic temples of nineteenth-century life, ringed the built-up area of Berlin as it existed at mid-century, but have almost all fallen victim to wartime destruction and post-war development (see fig. 4.2). The Hamburg Station of 1845–7 still survives as a building, but had been abandoned for rail purposes in 1884. Continuity is offered only by the Schlesischer Bahnhof, renamed Ostbahnhof by the GDR, which on the completion of extensive rebuilding operations is apparently to be Hauptbahnhof Berlin (Berlin Central Station). Improvements were also made to Berlin's waterways; the main link through the city provided by the Spree was supplemented by the parallel Landwehr Canal of 1845–50, and the Spandau Canal was completed in 1859 (see fig. 4.1).

At the same time Berlin was beginning its rise as a major industrial centre. The roots had already been established in the seventeenth and eighteenth centuries, stimulated by the demands of the army and the court. The army required the manufacture of cloth for uniforms, leather, military equipment, and arms (the latter industry located primarily in the vicinity of the fortress of Spandau). The requirements of the administration stimulated the development of paper-making and printing, while the court called for luxury products such as silk, carpets, porcelain (*Kgl. Porzellanmanufaktur* 1763) and objects in gold and silver. In this way branches of production were established which continue to be important for Berlin to this day, notably textiles, clothing and printing.

The early nineteenth century saw the addition to Berlin's industrial structure of the all-important branch of machine building. The state was influential in the early stages; in particular the establishment in 1803 of a royal iron foundry in the Invalidenstrasse, outside the Oranienburger Tor on the north-western fringe of the then built-up area of the city, helped to determine the sector Chausseestrasse–Wedding–Gesundbrunnen in which much heavy-industrial development was to take place in the first three-quarters of the century (see fig. 5.1). Local capitalists soon took up the running from the state. Near the royal iron foundry in 1821 Egells established a plant which is regarded as the point of origin for Berlin heavy machine building; close to it developed a range of similar establishments, some bearing names long prominent in Berlin industry, such as Borsig. The arrival of the railway was clearly important in the development of this type of industry, both as a means of transport and as a source of orders; when the Berlin–Potsdam line was built, rails and engines came from the United Kingdom, but other rolling stock was already locally built, and by 1841 Borsig was building steam locomotives. Another branch of Berlin industry having its origins in this north-western sector of the city is the manufacture of chemicals, especially pharmaceuticals, particularly associated with the name Schering. Then the establishment by Siemens and Halske near the Anhalter Bahnhof of a small workshop for the development of electric telegraphy announced the arrival of a further essential element in the branch structure of Berlin industry, the manufacture of electrical equipment (Vogel 1966: 119–21).

Economically significant also was the founding in 1821 of a technical school, from 1827 known as the Gewerbe-Institut, which together with the Building Academy (1799) and the Mining Academy (1770) provided the origins of the Technical College (Technical University from 1946), which is still at the service of the Berlin economy. Many of the great Berlin banks also owe their origins to this period. The municipal provision of gas (1847), the provision of filtered and piped water from the Spree by a British firm (1856), and the creation of a professional fire brigade (1851) were moves towards a modern urban organization. Berlin, like Prussia as a whole, was on the move towards economic changes of unprecedented extent, with social consequences that would require the intervention of an administrative machine of far greater complexity than that hitherto offered by an absolutist monarch and handful of honest Prussian bureaucrats.

1.3 Capital city on a world scale

The country that emerged victorious in 1870–1 was still predominantly rural, less industrialized even than France. The early 1870s mark a real change of pace in the development of Germany, a time of acceleration in the processes of modernization, industrialization, and urbanization. Not only had France been crushed; the external emergency had served to drive through the creation of a second Reich under Prussian dominance, uniting all German lands except Austria and German-speaking Switzerland. French reparations sparked off a hectic economic boom, the so-called *Gründerzeit*, the time of promoters. Germany changed from a country where, in about 1830, 80 per cent of the population earned a living from the land to one where, in 1895, the figure was barely 20 per cent. Total German industrial production overtook that of France in the 1870s, caught up with British output by the end of the century and clearly surpassed it by 1910, to become second only to the United States (Mann 1974: 332–5). The growth of industry was accompanied by a rapid surge in population, from 41 million in 1871 to 65 million in 1913 (area of 1914 Germany). A massive rural–urban migration ensured that this population increase was concentrated in the exploding cities, so that by 1910, 62 per cent of the population was urban and 38 per cent rural, almost exactly the reverse of the distribution of 1871 (Dickinson 1953: 192).

As we have seen, the Berlin that emerged as capital of the Second Reich had been just one town among many in the medieval period, and just one princely capital among many in the seventeenth and eighteenth centuries. This relatively late emergence ahead of numerous rivals meant that Berlin even in its imperial or National Socialist heyday never achieved the same degree of primacy within the national territory as did Paris or London. At least until 1933–45 the capitals of the surviving German states, such as Munich or Dresden, or the surviving free cities, such as Hamburg, retained some scope for independent decision-making in internal political matters. Economic decision-making was also widely dispersed among cities such as Munich, Frankfurt, Düsseldorf, or Hamburg, as well as Berlin. Nevertheless, Berlin could not fail to benefit from being the seat of the two most powerful political administrations in Germany – that of the Reich and that of Prussia – to which were added various administrative and official institutions of the Prussian province of Brandenburg.

Ministerial buildings clustered on the Wilhelmstrasse (over its length in East Berlin now known as Otto-Grotewohl-Strasse), which became synonymous with the centre of political power in imperial Germany in precisely the same way as Whitehall in the United Kingdom (see fig. 6.1). The greater European powers (United Kingdom, France, and the Soviet Union) had their embassies at the western end of the Unter den Linden, where even today the Soviet Union has a massive post-war building and the United Kingdom an extremely modest one (the site of the original embassy lying grassed over, but nominally in UK ownership, just by the Berlin Wall at the Brandenburg Gate). The ornate Reichstag building of 1884–94 on the Königsplatz lay rather remote on the fringe of the Tiergarten park, which also provided an overflow area for ministries (especially those concerned with national defence) and embassies (see fig. 7.4).

Organizations of economic control were attracted to the major centre of political power. The 1859–64 building of that essential institution of capitalism, the Stock Exchange (Börse) lay on the Spree opposite the Berlin Cathedral. Above all, this was the age of the banks, which clustered on the Behrenstrasse and neighbouring streets of the Friedrichstadt south of the Unter den Linden:

> Although most of them were founded before 1870, it was only now that they developed into organizations with thousands of employees, with marble palaces . . ., temples of the new God; there was the *Deutsche Bank*, the *Dresdener Bank*, the *Darmstädter Bank*, the *Disconto-Gesellschaft* and the *Berlin Handelsgesellschaft*. Their growth was stimulated by industry whose development they promoted; they financed industrial expansion, took part in setting up new firms and founded businesses themselves.
>
> (Mann 1974: 334)

At the head of the financial pyramid lay the 1857 Reichsbank at Hausvogteiplatz, initially founded with private capital, but with directors nominated by the Kaiser. Berlin overtook Frankfurt to become the principal banking centre of the Second Reich.

The process of central-area specialization that produced the government and banking quarters also produced a central concentration of retailing, which was attracted westwards out of the Berlin inner city towards the higher-quality residential areas. It settled down, also in the Friedrichstadt, along the east–west Leipziger Strasse:

> Its shops stretch in unbroken lines throughout the whole length of the street, from the Potsdamer-Platz to the Spittelmarkt, a distance of nearly a mile. . . . The *Warenhaus Wertheim*, which extends along

the N. side of the Leipziger-Strasse between the Wilhelm-Strasse and the Leipziger-Platz, was the first example of a great modern stores building. It was erected in 1897–1904.

(*Baedeker* 1936: 13–14)

Westwards the Leipziger Strasse terminated in the octagonal Leipziger Platz and the adjoining Potsdamer Platz, the latter a centre of cafés, restaurants, beer halls, and dance halls. With its constant flow of traffic from five intersecting main streets, and restless flows of people to and from the Potsdam main station, the Potsdamer Platz could be regarded as typical of the restless, crowded cosmopolitanism that was associated with the great capitalist cities, and which was regarded as particularly characteristic of Berlin. Great capitals are also the centres of information dissemination; the Friedrichstadt south of Leipziger Strasse was the home of press, printing, and publishing.

Associated with these new administrative and economic functions, and further stimulated by continuing industrial progress, the population of Berlin increased enormously. In 1871 the population of the area that was in 1920 to become Greater Berlin was still less than a million; by 1910 it was 3.7 million. Of the increase of 2.8 million, 68 per cent (1.9 million) was the result of immigration. Voluntary migrants being predominantly young adults, either with young children or about to found a family, it is not surprising that natural increase also made an appreciable contribution (32 per cent) to population growth (natural variation was to be negative throughout most of the twentieth century). With its rise to an imperial capital, Berlin became an almost inescapable goal for those wishing to rise to the very highest posts in administration, law, commerce, the academic sphere, or the arts. The great majority of migrants, however, took up lower-paid jobs in the expanding industrial sector, or in supporting tertiary occupations.

Migrants to Berlin came predominantly from Protestant areas. The immediate source lay in surrounding rural Brandenburg; the Prussian *Regierungsbezirke* Potsdam (excluding Berlin) and Frankfurt on Oder were the birthplace of about 60 per cent of the Berlin population in the second half of the nineteenth century. As the city expanded, it drew increasingly on rural areas yet further east and south-east: Silesia, Pomerania, Posen, East and West Prussia were the source of over a quarter of the Berlin population by the beginning of the present century.

The tradition of accepting immigrants from foreign countries was, at least for a time, maintained, but the Protestant Huguenots, Bohemians, and Salzburgers of the seventeenth and eighteenth centuries were in the nineteenth and early twentieth centuries replaced in terms of numerical preponderance by the Jews. The earliest Jewish population of the Brandenburg Mark had been driven out in 1573 following pogroms; the Jewish community of modern Berlin dates its origins to privileges conferred by the Great Elector in 1671, for which they had to make substantial payments. They were not allowed to participate in agriculture or manufacturing, and could not become civil servants. For most of them trade and money-lending were the only available occupations; a number became suppliers and bankers to the court. Some entered other self-employed occupations in the fields of medicine, the law, and the arts. It was not until 1850 that the Jews received at least formal equality before the law, and not until the foundation of the Reich of 1871 could they become civil servants. The early years of the Second Reich saw a rapid rise of the Berlin Jewish population, from about 47,000 in 1871 to 151,000 in 1910. By the fateful year 1933 the population acknowledging Jewish faith in Berlin was to rise to 160,000, a third of the total for Germany, but the population of Jewish descent was probably at least 250,000. Jews became prominent in all branches of Berlin life; in banking obviously (Mendelssohn), in press and publishing (Ullstein), politics (Lassalle, Marx), the world of learning (from Moses Mendelssohn onwards), hotels and restaurants (Kempinski), department stores (Tietz), music and the arts (Heine, Mendelssohn-Bartholdy, Max Liebermann), and science (Einstein). Their numbers included the geographer Friedrich Leyden (who died in Theresienstadt concentration camp, 30 January 1944), on whose book *Gross-Berlin: Geographie der Weltstadt* the present work draws significantly. Jewish participation was particularly 'visible' not only in finance but in occupations such as medicine and the law, something which may have contributed to the development of anti-Jewish sentiments.

The expansion of Berlin's population was paralleled by the expansion of the city, which largely took the form of the creation of an inner residential ring, often called the Wilhelmian ring, surrounding the Berlin inner city and its seventeenth-century and eighteenth-century extensions. The high-density development of the Wilhelmian ring predominantly took the form of monotonous five-storey apartment houses on a rectangular street grid, with narrow interior courts containing industrial and service establishments (see section 6.3).

This unprecedentedly large population needed service provision on a corresponding scale. The expansion of the city would hardly have been possible without the introduction of new forms of transport. A company for the provision of horse omnibuses was founded in 1865 and began the introduction of motor buses in 1905. Horse-drawn trams were also introduced in 1865 and became the dominant form of urban street transport until about the turn of the century, when they were replaced by electric trams. The S-Bahn system of rapid urban rail transportation was introduced in 1882; it was at about the turn of the century that the

S-Bahn, like the new U-Bahn (underground railway, subway), began to be turned over to electric traction.

Administratively, Berlin from the foundation of the Reich to 1914 was an anomaly, a city that gave witness to the triumph of the capitalist bourgeoisie but which was ultimately governed by an authoritarian and militarist imperial power. Ultimate authority for the government of the city lay in the Prussian Ministry of the Interior, after 1828 operating through the interior department of *Regierungsbezirk* Potsdam. The Prussian Police Authority (*Polizeipräsidium*) effectively acted as a kind of alternative urban administration. It was the *Polizeipräsidium* that initiated schemes for the supply of water and gas, for street lighting, drainage, street cleaning, rubbish removal, and the provision of public baths and wash-houses. The role of the police was particularly important in relation to the control of urban development. This is less surprising than it may appear to non-German readers, since it was common in the German-speaking lands for the 'Building Police' (*Baupolizei*) to exercise control over building lines along streets and over materials and construction methods in relation to building safety and fire prevention. It was at the request not of the municipality but of the police authority that James Hobrecht drew up the 1862 town plan that determined the nature of urban expansion in the Wilhelmian ring. This overriding of local government was advantageous to the degree that it allowed uniform planning to extend beyond the political boundary of the city of Berlin as it existed before 1920.

In the last quarter of the nineteenth century there was a gradual increase in municipal attempts to meet the social needs of an exploding population, helped by the provision for the first time of an adequate basis of municipal taxation. In 1873 the British water company was bought out, and over the next decades new water works were created on the Müggelsee and the Tegeler See. In 1875 the city also took over the administration of roads and bridges, except those in the vicinity of royal palaces and parks. Hospitals were established, especially in the densely populated inner districts such as Friedrichshain, Wedding, and Kreuzberg, assisted after 1883 by the introduction of a national system of health insurance. To improve sanitary conditions a new system of sewers was initiated in the 1870s, a central slaughter-house opened in 1881, and central markets in 1881. The first electric-arc street lighting was installed in the Leipziger Strasse and the Potsdamer Platz in 1882, and in 1884 the predecessor of the later BEWAG, the Berlin electricity works, was established. Some hundreds of schools were also built in this period, and facilities for higher education rapidly extended (the Technical College, later Technical University, in Charlottenburg, 1879). To house the increasing number of municipal officials the 'Red Town Hall' (*Rotes Rathaus*) of 1861–70 (the reference is to the bricks from which it is constructed, not to the municipal politics of the time) was supplemented by the 1902–11 *Stadthaus*.

A range of political and administrative problems remained unresolved in the Berlin of the 1871 Reich. The overlapping competences of national and municipal administrations remained a problem to the end, exacerbated by the fact that the Berlin bourgeoisie remained predominantly Liberal politically, electing 'Progressive' or Liberal representatives to the city council and even to the Prussian and Reich parliaments (in 1867 declining to elect Bismarck and five Prussian generals to the latter body). At the same time Berlin, as a great industrial city, witnessed the rise of the working-class movement, initially under Lassalle, who was to be attacked by Karl Marx (a student at the University of Berlin from 1836 to 1841). In 1868 the Berlin Congress of the working-class movement led to the formation of the first general trade unions. The leaders of the Social Democrats under the Reich, such as William Liebknecht and August Bebel, had to endure periods of imprisonment; nevertheless, the cause went forward. Berlin elected its first two Social Democratic Party (SPD) representatives to the Reich parliament in 1877, and when the party first participated in Berlin municipal elections, in 1883, it secured five seats. There is plentiful suggestion here of social tensions to come.

From the point of view of the future of the city, the biggest piece of unfinished business was the provision of a spatial unit of administration appropriate to a great capital, centre of an agglomeration of population approaching 4 million by the outbreak of the First World War. Berlin at the end of the eighteenth century was extremely small in territorial extent – only 1,343 hectares. When in the nineteenth century the built-up area of the city began to expand over these narrow political limits, it was inevitable that the question of the appropriate boundary of the city against the surrounding rural *Kreise* (administrative districts) of Brandenburg Province would become a subject of debate.

There was no unanimity over the best course to pursue. In the first half of the century, the authorities of the neighbouring *Kreise* Teltow and Niederbarnim accused Berlin with some justification of wanting to take over prosperous districts to the west and south-west, such as Charlottenburg, but not industrial quarters, such as Wedding. There was at one time a proposal for the creation of a separate circular *Kreis* around Berlin, embracing all the suburbs somewhat on the model of the Département de la Seine around the city of Paris. In 1860, however, the government of Prussia decided that Berlin should be extended by adding precisely those quarters which were either taken up by substantially working-class populations, or which were about to be so, such as Wedding, Moabit, Gesundbrunnen, and the northern parts of Tempelhof and Schöneberg. The result was to increase the area of Berlin from 3,510 hectares to 5,923 hectares. There was a small addition in 1878 from Lichtenberg on the eastern side of Berlin to permit the building of the central slaughter-house, but apart from that the major administrative addition of the later nineteenth century was the Tiergarten park, transferred from the *Kreis* of Teltow in 1881. The richer western suburbs, most notably Charlottenburg, retained their independence.

The *Gemeinden* (civil parishes) around Berlin continued to grow in population as the built-up area expanded. Six of them – Charlottenburg, Lichtenberg, Neukölln, Schöneberg, Spandau, and Wilmersdorf – were formally recognized as towns; some, such as Charlottenburg, attaining the category of *Grossstadt*, given to towns of 100,000 population or above. The determination of the suburban centres to maintain their separate identity was demonstrated by the construction in 1890–1916 of no fewer than seventeen imposing town halls, which still dominate quarters of the present city such as Charlottenburg, Schöneberg, Wilmersdorf, or Pankow.

In the 1890s, when it seemed as if a take-over by Berlin of the surrounding suburbs had become extremely unlikely, an alternative solution began to be put forward. This was the *Zweckverband*, an association between the city of Berlin and surrounding areas for specific administrative purposes (Hofmeister 1975a). A law to put this into effect was passed by the Prussian parliament in 1911, and came into force in 1912. It covered the city of Berlin, the suburban towns and the *Kreise* of Teltow and Niederbarnim, in all nearly 400 distinct *Gemeinden*. The *Zweckverband* was intended to have a co-ordinating role, particularly in the fields of urban planning, recreational space, and urban transport. It has been suggested that the very wide territorial limits of the *Zweckverband* reflected the desire of the Prussian government to balance politically 'unruly' Berlin by including not only the higher-income suburbs but the deeply conservative populations of the surrounding rural areas.

Some co-ordination was certainly necessary; the Berlin region had seventeen separate water supply systems, forty-three gas and electricity supply undertakings, and nearly sixty separate sewerage systems. Each suburban town or larger urbanized *Gemeinde* characteristically developed its own independent sewerage system, complete with an expensive individual pipeline out to its own separate sewage farm, where the effluent was spread over carefully contoured fields (*Rieselfelder*). Public transport was another area obviously crying out for co-ordination; the *Zweckverband* discovered that there were no fewer than 150 different agreements between its various members and public-transport enterprises. Some real progress was made in respect of environment and recreation through the purchase from the Prussian state of the Grunewald, the Tegel, Grünau, and Köpenick forests, and part of the Potsdam forest. The result was to preserve from speculative development an area of over 10,000 hectares of forest, agricultural land, and water, most of which still remains as a precious recreational resource for the people of the city. But the First World War was at hand, so that little further progress was made, except perhaps that a certain degree of drawing together of the population in the face of wartime adversity may have done something to induce a feeling among all inhabitants of belonging to a 'Greater Berlin'. Certainly, another solution to the problem of Berlin's political geography was to emerge after the war.

1.4 Greater Berlin 1920–45

Berlin entered the post-war period with revolution and violence. On 9 November 1918 the Republic was proclaimed by Philipp Scheidemann before the Reichstag and by Karl Liebknecht from the balcony of the Berlin Palace; Kaiser Wilhelm II went into exile. In the course of the left-wing Spartakus attempt to seize power in 1919, Karl Liebknecht and Rosa Luxemburg were apprehended and shot in the Tiergarten by proto-fascist *Freikorps* members, their bodies being thrown into the adjoining Landwehr Canal. In 1920 it was the turn of the counter-revolutionary Kapp-Putsch, frustrated by the solidity of the Berlin trade unionists. Although the Weimar Republic settled down to a sort of normality, throughout the period until 1933 there was intermittent street fighting between right and left in Berlin, and political murder was far from unknown. The seamier side of Berlin life at this time has perhaps been made too well known to English-language readers through the novels of Christopher Isherwood. Yet in all fields but politics the inter-war years, and in particular the 1920s, came to be looked back on as a 'golden age'. In spite of the 1923 inflation and the world economic crisis of the early 1930s, Berlin made a surprising degree of economic and social progress, even if in the end victory went to the forces of reaction.

One of the achievements of Weimar Germany was the creation in 1920 of a unified Greater Berlin out of the urban civil parishes of Berlin, Charlottenburg, Köpenick, Lichtenberg, Neukölln, Schöneberg, Spandau and Wilmersdorf, as well as fifty-nine rural civil parishes and twenty-seven rural estates (*Gutsbezirke*). The area of Berlin thereby increased thirteen-fold, from 6,572 hectares to 87,810 hectares, and the number of inhabitants was approximately doubled, from 1.9 million inhabitants to 3.8 million (Table 1.1). These boundaries of the city were, with some minor alterations, retained in 1945, a fact of major importance given the exceptional occupation regime accorded the city at this time (see fig. 2.1). The new boundaries did not fully accord with the actual pattern of settlement. They included extensive areas of land that were not built up, not only forests and lakes but farmland and villages (subsequently to become precious resources for the 'island' city of West Berlin, and in not much lesser degree for East Berlin). On the other hand, tentacles of industrial and urban development extended beyond the new boundary along the railways, for example north-westwards to Oranienburg, or south-eastwards to Königs Wusterhausen. The most important exclusion was, however, the town of Potsdam, left just beyond the boundary to the south-west, although closely linked to Berlin not only by history but by intense daily commuter flows. The *Kreise* surrounding Berlin, which had already lost heavily in population and tax income from the creation of Greater Berlin in 1920, were able to fend off any further significant expansion of the city territory, while after 1933 the National Socialist leaders seem to have been more concerned with their grandiose plans for restructuring central Berlin than with changing its outer limits. Some limited co-operation between Berlin and its surrounding tributary area was introduced by

Table 1.1 *Berlin: growth of area and population*

Year	Historical event	Area in hectares	Population
1648	End of Thirty Years' War	83	6,000
1709	Amalgamation of five towns	626	61,000
(1712)			
1815	Urban reforms	1,400	193,000
1848	Year of revolutions	3,510	419,000
(1850)			
1861	Inclusion of Moabit & Wedding	5,920	428,500
1912	Foundation of Zweckverband	6,572	2,072,000
	(In area of future Greater Berlin)		3,734,400
1920	Foundation of Greater Berlin	87,810	3,803,300
1943	Maximum population	88,370	4,489,700
1945	End Second World War	88,994	2,807,400
1961	Building of Berlin Wall	88,308	3,252,691
	West Berlin	48,008	2,197,408
	East Berlin	40,300	1,055,283
1985	Mid-1980s	88,308	3,075,670
	West Berlin	48,008	1,860,084
	East Berlin	40,300	1,215,586
1987	750th Anniversary (approx.)	88,308	3,090,000

Source: Hofmeister (1975a), with later additions.

the formation in 1929 of a regional-planning association, *Landes-planungsverband Brandenburg-Mitte* (from 1937 to 1945 absorbed into *Landesplanungsgemeinschaft Brandenburg*) (Pfannschmidt 1971; Engeli 1986).

The new Greater Berlin continued to be administered as a unified city by the *Magistrat* (city council) installed in the nineteenth-century city hall in the inner city. The Berlin bear, which had first appeared on the seal of the city in 1280, continued as the Greater Berlin symbol. To provide for a certain degree of administrative decentralization the *Magistrat* was represented in each of twenty *Verwaltungsbezirke* (administrative districts) into which the city territory was divided by a local office (*Bezirksamt*). Typically, these district offices utilized the grandiose town halls built in the last years of suburban independence. The exact division of competence between *Magistrat* and the *Bezirk* offices was never properly worked out, so that a certain amount of friction persisted until after the Second World War. The districts were of disparate composition, reflecting the uneven settlement structure of the area that was to become Greater Berlin. Apart from *Bezirk* Mitte, enclosing the heart of the city, five districts were based on formerly independent towns (Spandau, with urban status from the thirteenth century; Charlottenburg from the early eighteenth century; and Neukölln, Schöneberg, and Wilmersdorf from about 1900). Tiergarten, Wedding, and Kreuzberg and Prenzlauer Berg were based on sections of the fringe of pre-1920 Berlin, while outer districts like Steglitz, Zehlendorf, or Weissensee were cobbled together from mixtures of old villages and recent suburbs. The former small country town of Köpenick, dating back to medieval times, was made the centre of a large *Bezirk* containing a mixture of industry, housing, forests, and lakes. Populations also varied greatly, as they do to this day, ranging from an initial 33,000 in Zehlendorf to 366,000 in Kreuzberg. Social composition also varied, from the dense-packed working-class *Miets-kasernen* of Kreuzberg or Prenzlauer Berg to the suburban villas of Zehlendorf.

The creation of a unified Greater Berlin was followed by the unification of a range of services. A host of small suppliers was replaced by the founding of three city-owned companies for the supply of water, gas, and electricity (the Klingenberg power station dates from 1926). In 1926 the city took over the U-Bahn, and in 1928 the Berliner Verkehrsgesellschaft (BVG) provided a single organization uniting all bus, tram, and U-Bahn services in the city.

The inter-war years, but especially the 1920s, was a period not only of population growth but of urban explosion out of the dense-packed Berlin that had been created by 1914. There was a continuation of the process of decentralization of the larger industrial firms to new sites on the railways and waterways leading away from the urban core, that had begun at the end of the nineteenth century. Sometimes, as at Siemensstadt, there were accompanying housing estates for the workers. The electrification and extension of the S-Bahn system and the construction of the first U-Bahn lines permitted the growth of higher-income suburbs, like those fringing the Grunewald, or 'garden cities' such as Frohnau. The most impressive achievement, however, was the construction, mainly in the eight years from 1924 to 1932 that separated the inflation period and the world economic crisis, of a series of publicly supported housing projects for people of moderate means; at the peak of development in 1927, nearly 27,000 dwellings a year were being produced (see section 6.4.3). The National Socialists did nothing like this; Hitler was more concerned with prestige projects, such as the Olympic Stadium, which was at least finished, or the grand axis that was to run southwards across the city from the Reichstag to a new South Station near the Tempelhof Central Airport, on which work had scarcely begun before being brought to a halt by the Second World War.

The inter-war period was also one in which Berlin appeared to be making further advances towards a position of absolute political, economic, social, artistic, and intellectual predominance in the German lands. It is true that internal power had to some extent to be shared with other German cities, but the Wilhelmstrasse had undisputed dominance at least over foreign policy.

Economically, Berlin was pre-eminent in banking and finance; if of 290 banks in the Reich in 1939, Berlin had the head office of 'only' seventy-four, they were important enough to control two-thirds of the country's banking capital. The Stock Exchange, on the other hand, had already before 1939 come to be regarded as marginally less important than that of Frankfurt on Main. Although Berlin was only one of Germany's major industrial centres, the city nevertheless contained the headquarters of some of Germany's greatest industrial firms, with such famous names as Siemens, AEG, Schering, and Sarotti.

In the field of the arts, Berlin became a goal for talent from all over the German-speaking lands. With forty-five stages it was the undoubted theatrical metropolis, associated with such names as Max Reinhardt. Its opera houses (Erich Kleiber, Otto Klemperer, Bruno Walter) and symphony orchestras (Wilhelm Fürtwängler) were of international renown. Until 1933 Paul Hindemith taught in the High School for Music. Berlin was also the centre of a growing film industry (the principal studios were established just across the city boundary in Potsdam-Babelsberg). In addition to the superb museums, temporary exhibitions and art auctions drew an international attendance. Max Liebermann was president of the Academy of Fine Art, Heinrich Mann headed the Academy of Literature.

In the world of learning, the vast Friedrich-Wilhelm University and the Technical College in Charlottenburg at least equalled their rivals in other German cities, attracting teachers and students from all over the German-speaking lands. In addition, university institutes, laboratories of the Kaiser Wilhelm Research Association and government laboratories were established in the West Berlin suburb of Dahlem, a fact of subsequent importance in the post-1945 divided city. Here, in 1939, the successful fission of uranium by Otto Hahn and Fritz Strassmann presaged a new scientific age.

The city was absolutely predominant in the dissemination of information. It was Germany's biggest centre for newspaper and magazine publication, with in 1928 no fewer than 147 daily and weekly publications (Vogel 1966). After the First World War Berlin also developed a leading position in radio broadcasting; the 1926 Funkturm (Radio-tower) in Charlottenburg is still a feature of the West Berlin skyline. Hans Poelzig's 1931 Haus des Rundfunks (Broadcasting House), like its equivalent in London, is now something of a monument of early twentieth-century architecture. The adjacent halls of the Exhibition Centre date from much the same period, and point to another aspect of Berlin's growing national and international significance at this period.

Berlin continued to be a major centre of retailing and entertainment. A feature of this time was the tendency for the establishment of a second centre of retailing, fashionable cafés, and entertainment facilities in the west of the city in the neighbourhood of the Kaiser Wilhelm Memorial Church, of which a contemporary account wrote:

> Its silhouette is lit up at night by the glare of the illuminated advertising signs. From the late afternoon until long after midnight it is encircled by an unbroken stream of traffic. Cinemas, dance halls, cabarets, restaurants, smart shops, and miscellaneous places of entertainment are grouped round the church and have overflowed into the neighbouring streets, notably the wide Kurfürsten-Damm.
>
> (*Baedeker* 1936: 16)

The emergence of this second urban core was to be of great importance in the spatial reorganization of the city after 1945.

Berlin's growing centrality was accompanied by further improvement in its system of communications. It became the principal centre of air transport in Central Europe, its importance symbolized by the 1936–9 terminal building of the Tempelhof Central Airport, which had what was at the time the advantage of being created on former military land only 3 km from the city centre. The Avus motor road (*Automobil-Verkehrs- und Übungsstrasse*) running south-west from the city presaged the building of Hitler's autobahn system, for which the ring motorway around Berlin, incomplete until well after the Second World War, was conceived as the centre.

In the National Socialist period Berlin lost much of its brilliance through the emigration, imprisonment, or otherwise silencing of many of its leading figures, but gained in internal political significance through the pervasive intervention of the new order in all aspects of national life; one has only to think, for an example, of the tight control of Goebbels's vast Propaganda Ministry over all aspects of information dissemination throughout the Third Reich. For the space of a few years, Berlin was the capital of an empire stretching from the Channel to the Caucasus, from the North Cape almost to the Nile, but all was to come to dust under the pounding of Anglo-American bombs and direct assault by the Red Army.

2 | Berlin divided

2.1 Occupation

On 16 April 1945 Generals Zhukov and Konev launched the final assault on Berlin. The Soviet forces had the benefit of an enormous superiority of men and equipment, but it was not until 2 May, after days of bitter street fighting, that the remnants of the German garrison capitulated. Two days before, on 30 April, Hitler had killed himself in his Bunker in the grounds of the Chancellery in the Wilhelmstrasse. As early as 28 April, well before the fighting was at an end, Colonel-General Bersarin issued an order announcing that he had been appointed commander-in-chief of the occupation troops and chief commandant of the city of Berlin, and that all administrative and political power lay in his hands (GDR Ministry of Foreign Affairs 1964, Documents). The post-war political process that was to end in a divided Berlin had begun. The drama that was to unfold on the stage offered by Berlin, however, was only partially related to imperatives emerging within the city itself. It was to a much larger extent dictated by the development of relations between two parts of a divided Germany, and even more to decisions and events at international level. There are varied opinions as to responsibility for the course taken by events and, in particular, markedly divergent 'western' and 'eastern' official versions.

The decisions that were to lead to the division of Germany and of Berlin had been taken by the wartime Allies, the United States, the United Kingdom, and the Soviet Union, before the end of the war in Europe. In contrast to the armistice that had terminated the First World War, which had left the German state in being, this time the demand was for unconditional surrender. The German state was to cease to exist, which meant that the victors were obliged to provide for some kind of occupation administration, at least in the short run. It was decided that the Soviet Union would be allowed to occupy the north of East Prussia (with promise of support from the western Allies for permanent inclusion at an eventual peace treaty). Poland was to annex provisionally ('until the conclusion of a Peace Treaty') the remaining territory east of the rivers Oder and Western Neisse. The Poles proceeded to expel the German-speaking population, thus making irreversible the provisional arrangement of the wartime agreements. Arrangements regarding the administration of the remaining three-quarters of Germany stem from the work of a European Advisory Commission, set up at the Moscow conference in autumn 1943. The recommendations of the Commission, enshrined in the decisions of the various inter-Allied meetings of 1944–5, were marked by considerable vagueness about the nature of a future administration, but it has been argued that this vagueness was a necessity of the time: 'the preservation of the Alliance . . . depended upon the postponement of decision' (Windsor 1963: 19). Such a postponement unfortunately resulted in plentiful opportunities for subsequent differences of interpretation.

The European Advisory Commission recommended that Germany was to be divided into zones of occupation. Lines of division into three zones, devised in 1943 by a British cabinet committee under Clement Attlee, were adopted virtually unchanged. It appears that the division adopted was not greatly to the taste of the Americans, whose diplomatic effort at the time was, however, rendered ineffectual by a division of aims between the President, the military and the State Department, so that the version favoured by the United Kingdom and the Soviet Union prevailed (Smith 1963).

It was agreed at the Yalta Conference in February 1945 that a fourth zone should be created for the French out of the UK and US zones, the Soviet zone remaining unchanged. The formulation of the European Advisory Commission was taken up almost word for word in the Potsdam Agreement of 2 August 1945, section 111.A.1:

> In accordance with the Agreement on Control Machinery in Germany, supreme authority in Germany is exercised, on instructions from their respective governments, by the Commanders-in-Chief of the armed forces of the United States of America, the United Kingdom, the Union of Soviet Socialist Republics, and the French Republic, each in his own zone of occupation, and also jointly, in matters affecting Germany as a whole, in their capacity as members of the Control Council.
>
> (GDR Ministry of Foreign Affairs 1964b: 39)

Already, then, we have the somewhat difficult position of the commanders-in-chief each having absolute power in his own zone, but expected to consult with each other in securing some kind of administration for a Germany that all of the Allies expected to be treated as a whole. One reason for a unified occupation administration, particularly important for the war-torn Soviet Union, was to allow a joint resolution of the question of reparations payments. It was certainly envisaged at the various wartime meetings that a future political settlement would involve the dismemberment of Germany into a number of separate states, but nobody appears to have anticipated that the zones would freeze into a permanent political structure (Windsor 1963).

Beyond that, there was the question of Berlin, which was quite clearly embedded geographically within the territory over which the Soviet commander-in-chief was deemed to exercise supreme authority, but which was subjected to special provisions. The city would be jointly occupied and would be the seat of the Control Council, symbolizing the intention of the Allies to work together for a common political solution. According to the 'Record of the Agreement of 12 September 1944, as revised on 14 November 1944 and 26 July 1945':

1. Germany, within the frontiers as they were on 31 December 1937, shall, for the purposes of occupation, be divided into three [subsequently four] zones, one to be allotted to each power *and a specific area of Berlin, to be jointly occupied by the four powers.*
2. [. . .]

The area of Berlin (meaning the territory of 'Greater Berlin' as laid down in the law of 27 April 1920) shall be *jointly occupied* by the forces of the USSR, the USA, the United Kingdom and the French Republic, to be designated by the responsible Commanders-in-Chief. *The territory of 'Greater Berlin' shall be divided into three* [subsequently four] *parts for this purpose*:

The north-eastern part of 'Greater Berlin' (the districts of Pankow, Prenzlauer Berg, Mitte, Weissensee, Friedrichshain, Lichtenberg, Treptow and Köpenick) shall be occupied by the forces of the USSR.

The north-western part of 'Greater Berlin' (the districts of Reinickendorf, Wedding, Tiergarten, Charlottenburg, Spandau and Wilmersdorf) shall be occupied by the United Kingdom [a decision of the Control Council dated 30 July 1945 subsequently reallocated the districts of Reinickendorf and Wedding to the French].

The southern part of 'Greater Berlin' (the districts of Zehlendorf, Steglitz, Schöneberg, Kreuzberg, Tempelhof and Neukölln) shall be occupied by the forces of the United States of America.

(GDR Ministry of Foreign Affairs 1964b: 25–6)

Article 3 of the text of the 'Agreement on the Machinery of Control in Germany of 14 November 1944 in the revised version of the Agreement of 1 May 1945' makes it clear that one of the functions of

Figure 2.1 Greater Berlin and its administrative divisions 1920–c.1980

the Control Council for Germany would be the direction of the administration of Greater Berlin (fig. 2.1). Article 7 of the same Agreement gives some additional information with regard to Berlin:

a) An Inter-Allied Authority (Russian: Kommandatura) consisting of four commandants, one of each occupying power and appointed by their respective commanders-in-chief, will be established for the *joint direction of the administration* of the 'Greater Berlin' area. The Inter-Allied Authority will be directed by a chief commandant to be designated in rotation by each of the allied commanders in chief.

b) A technical staff, formed by representatives of each of the four powers, shall be attached to the Inter-Allied Authority; in its structure it shall correspond to its functions of supervising and controlling the activities of the local German organs responsible for the city administration of 'Greater Berlin'.

c) The Inter-Allied Authority shall operate under the general direction of the Control Council and receive instructions from the co-ordinating committee.

Article 10 of the same agreement further states:

The allied authorities for the control and administration of Germany described above shall operate during the initial period of occupation immediately following German surrender, i.e. during the period when

Germany is carrying out the basic requirements of unconditional surrender.

(GDR Ministry of Foreign Affairs 1964b: 27–8)

(The extracts are taken from the German translation of 'Collection of Valid Agreements, Treaties and Conventions concluded between USSR and Foreign States', edited by the Ministry of Foreign Affairs of the USSR, 1955, v. 11, pp. 55 and 62; original in Russian. Reprinted in GDR Ministry of Foreign Affairs 1964b. Italics added.)

It was on the rather slender and contradictory basis of these agreements that the first troops of the western Allies entered Berlin on 1 July 1945. Retrospectively, it is incredible that they did so without explicit agreements securing freedom of access to the city for themselves and also in respect of the civilian population of their sectors and their necessary supplies. The lack of a specific access agreement is all the more surprising in view of the care taken by the United States to extract cast-iron guarantees from the British regarding freedom of access to their zone from their port-enclaves Bremen and Bremerhaven, and similar undertakings with regard to the French zone (Smith 1963). It has to be remembered that the agreements were drawn up at a time when the Americans were more inclined to be suspicious of the motives of the 'imperialist' and 'colonialist' Britain and France than of their Soviet ally. There was also a marked disinclination to do anything to irritate the country that was bearing the main brunt of the war in Europe and which the Americans hoped to enlist for what was seen as a protracted further campaign against Japan. It was a honeymoon period, when a continuing unity of purpose of all the Allies was assumed; the discords to come were not foreseen. If Germany and Berlin were to be administered as a unity under the supreme authority of the Control Council, special access arrangements were clearly unnecessary. In any event, the occupation arrangements were temporary, pending a peace treaty with a unified Germany that could not be far distant in time. The matter of access was, indeed, raised at military level just before the forces of the western Allies moved into Berlin, but the Soviet commander-in-chief, Marshal Zhukov, was willing to offer only a verbal agreement to the use of a single land route, the Helmstedt–Berlin autobahn and one air corridor Berlin–Magdeburg (diverging then to Hanover and Frankfurt), and this only as a privilege, not as a right (Smith 1963).

The momentum of the advance of the western Allies in the closing days of the war led to their occupation of half of the territory of the future GDR, containing two-thirds of its population. The Americans captured Magdeburg on 11 April and Leipzig the following day, then waited for two weeks on the Elbe at Torgau for the Soviet army to reach the river. Overruling Churchill's protests, the Americans vetoed an advance on Berlin, which was left to the USSR. The British and Canadians meanwhile advanced across the lower Elbe as far as Wismar. At the same time that they sent forces to Berlin the western Allies withdrew their forces from these advanced positions, which in the euphoria of wartime collaboration they declined to retain as bargaining counters, for example to secure an adequate access agreement for Berlin. The Soviet Union and the GDR were subsequently to deny that the occupation of West Berlin was in any way a *quid pro quo* for these withdrawals.

The growing tensions that led to the division of Germany and to the blockade of Berlin have greatly varying interpretations. A fairly standard 'western' view would be that the western Allies arrived in Germany with no fixed objectives other than a general goodwill towards the Soviet Union and a willingness to operate the occupation agreements in good faith. Such partial plans as had been put forward for discussion, such as the Morgenthau plan for turning a de-industrialized Germany into a nation of peasant farmers, rapidly revealed their impractical nature. There was an urgent need to give the zones a stable political and economic structure, so that they might at least produce enough exports to provide for necessary imports of food for a highly urbanized population. These practical needs were so urgent that it was impossible to await the achievement of an all-German solution through agreement with a Soviet Union that had proved demanding and dilatory as a negotiating partner. According to this version, the Soviet Union is perceived as having indeed had the aim of a united Germany, but one with a 'popular front' type of government that would be dominated by a coalition of the Socialist and Communist parties, with the latter as senior partner. A second objective, likely somewhat to impede the achievement of the first, was that this united Germany, which would include the industrial Ruhr, would supply reparations from current production to do something to make up for the appalling damage inflicted by Germany upon the Soviet Union. This revived Germany would, of course, be neutralized, and after the withdrawal of the occupation forces the Soviet Union would have an effective veto over governments that did not meet its approval, rather on the model of Finland (Mander 1962: 39–40).

The Soviet Union was, however, unable to obtain the co-operation and ultimate merger of the Communists and Social Democrats in those parts of Germany over which it had no direct control, and signally failed to overcome the vehement anti-Communism of the Social Democratic Party (SPD) of Schumacher in the Western occupation zones and Ernst Reuter in Berlin, thus rendering nugatory this political programme. Soviet attempts to extract reparations from the western zones of occupation and generally to milk their derelict economies with the help of Occupation Marks fresh off the printing press infuriated the western Allies, who were forced as a result to divert their own resources to provide relief supplies to fend off starvation and disease (this at a time when the population of the United Kingdom was still subjected to strict food rationing). The western Allies felt obliged to resort to economic measures affecting their own zones only, culminating in the 1948 introduction of a separate western currency. The argument runs that the Soviet Union, failing to achieve its political aims on the scale of a unified Germany, settled for the second best of full political control over the Soviet zone, thus sabotaging the inter-Allied agreements and causing the division of Germany. At this stage the western sectors of Berlin became, in the Soviet view, an irritant and an irrelevancy, to be removed as rapidly as possible.

Of course, an 'eastern' view of developments would be distinctly different. It would be argued that the Yalta and Potsdam agreements on the post-war treatment of Germany had been unilaterally breached by the western Allies through the adoption of separate political arrangements for the western zones, with the aggressive purpose of producing a re-armed Western Germany, as a hostile act against the Soviet Union. Having destroyed the main agreements, the western Allies automatically surrendered their rights under subsidiary agreements, such as those initially allowing their presence in West Berlin, which had been quite unacceptably developed into a base for western militaristic, fascist, and espionage activities (GDR Ministry of Foreign Affairs 1961a: 4–19).

The western Allies arrived in a Berlin that had already been administratively reorganized on the general model applied in the Soviet zone; they were not entirely pleased with what they found, but their first reaction was to try to operate the system as it existed. The 'Ulbricht Group' of German Communists that had lived out both the Hitler period and the Stalinist purges in the Soviet Union left Moscow on 30 April 1945. On reaching Berlin they were charged by the Soviet military government with setting up a civil administration, with not more than half the posts in Communist hands. Wolfgang Leonhard, who was involved in this process, has described how, in the middle-class district of Wilmersdorf, he found a former member of the old conservative Deutsche Volkspartei to be mayor, but appointed Communists to the key posts of deputy mayor, chief of police and head of education (Leonhard 1955). Similarly, as early as 17 May the Soviet occupying power had nominated a new city council (*Magistrat*) with the non-political civil engineer Dr Arthur Werner as governing mayor (*Oberbürgermeister*) and a range of other non-political specialists as members. However, the deputy mayor and about half of the members were Communists, and the key position of police president was occupied by Paul Markgraf, a German officer captured at Stalingrad, who had identified with the Soviet occupiers. A beginning was made with the setting up of a unified trade-union system, subsuming all political tendencies under Communist leadership, an anti-fascist judiciary, a 'Democratic Women's League' and a 'Cultural League for the Democratic Renewal of Germany'. Centralized institutions (located in the future Soviet sector) were also set up to control the banking and social-insurance systems, thus tightening the Soviet grip on the economy as a whole.

All these bodies had been put in place before the western Allies arrived and, together with the Soviet Military Administration (SMA), had proved effective in overcoming the immediate problems of destruction, hunger, and disease threatening the city in the first months after surrender. However, at the meeting of the Allied Kommandatura on 11 July, the four occupying powers further agreed that all orders issued by the Soviet occupying power or by the German authorities under their control prior to the arrival of the western Allies were to remain in force until further notice (GDR Ministry of Foreign Affairs 1964b, Document 13). Since changes could only be made by the unanimous decision of the Allied Kommandatura, this decision meant that the western Allies had not merely to accept the *fait accompli* but would find it difficult in the future to amend the composition of local government bodies, police, political parties, and trade unions, not only at city level but even within the districts (*Bezirke*) of their sectors of occupation. The city was in this way cast into a Soviet-dictated mould; it was clear that it was the Soviet intention that the Western military occupation was intended to be symbolic, without consequences for the political and economic organization of the city.

The western sectors had also in the weeks before the arrival of the western Allies been subjected to particularly heavy dismantling of equipment in manufacturing industry, in public utilities, and in transport, for shipment on reparations account to the Soviet Union. In addition to direct war losses of about 23 per cent, dismantling has been estimated to have removed a further 53 per cent of pre-war industrial capacity, or 70 per cent of that remaining at the end of the war. To quote individual branches, only 9 per cent of machine-building capacity and 15 per cent of electrical-equipment capacity survived. While the hand of the Soviet Union on East Berlin and the Soviet zone of occupation was not exactly light so far as dismantling was concerned, it was lighter than in West Berlin, presumably in the hope of receiving a continuing flow of reparations out of current production, as was indeed to be the case.

These and other internal causes of friction might have been resolved had not the city been swept up into conflicts on a German, European, and world scale. Seen from the narrower viewpoint of the western occupiers of Berlin, the pattern was one of an initial determination to seek accord with the Soviet occupiers being slowly replaced by irritation at the increasing barriers created by the SMA against any unified administration of the city, if it could not be on Soviet terms. It became clear that positive results were not to be expected from the Allied Kommandatura, nominally providing a unified Berlin administration, the SMA preferring to work through the *Magistrat*, the members of which they had nominated. The SMA could also bypass the Kommandatura by claiming that Berlin, being an integral part of the Soviet zone, was subject to decrees relating to that zone. Another ploy was the claim that the Soviet zone was the unique inheritor of the property of the former state of Prussia. This claim came to the fore in relation to the reopened Berlin University, which the Soviet military administration claimed as uniquely related to the Soviet zone, and outside the powers of the *Magistrat* or the Kommandatura.

The western Allies resented the uncooperative behaviour of Communist mayors and officials in their sectors, and a number were eventually removed. There were disagreements over the exclusive Soviet control of press and radio, while the Markgraf police showed itself to be far from impartial. On the positive side was the agreement initiated in November 1945 and finally concluded in early 1946 that there should be three air corridors to Berlin, from Hamburg, Hanover, and Frankfurt, and that all air traffic in a radius of 20 km from central Berlin should be controlled by a Four-Power Berlin Air Safety Centre. The Air Safety Agreement has proved to be the most durable example of Power co-operation in Berlin.

2.2 Division and blockade

In view of the reluctance of the western Allies to provide anything other than unquestioned support of the Soviet position throughout most of the first year of their occupation, it is ironic that it was a crisis within the coalition of political parties created by the SMA from members of the former German enemy that provoked a crisis in relations. In Berlin the rank and file of the reconstituted SPD refused to go along with a merger of the Socialist and Communist parties to form the Socialist Unity Party (SED) that was being energetically pushed by the SMA with the notion that it would obtain political leadership not only in the Soviet zone but throughout Berlin and in the western zones of Germany. This proposal for a unified working-class party under Communist leadership had the support of prominent SPD leaders in the Soviet zone, notably Otto Grotewohl; its rejection in a referendum of party members in the western sectors of Berlin, where alone voting could take place freely, was a major blow to Soviet plans:

> This was the first decisive western victory in the political battle for Berlin. The effect was that the concealed bid of the Russian and German Communists was now transferred to an open conflict among the Allies, which hastened the development of the Cold War over the whole of Germany.
>
> (Windsor 1963: 70)

The *Magistrat* elections of 20 October 1946 had a yet more dramatic result; even in the Soviet sector the SPD won more votes than the SED in every borough, and a non-Communist coalition was installed. This development is regarded by many as the event that, more than any other, made it inevitable that the Soviet Union would respond by endeavouring to gain its ends in Berlin by other than normal political means. Seen from another point of view:

> Under the cover of Western occupation, Berlin was to be turned into a 'bridge-head' in the heart of Eastern Germany in order to disturb further construction there. From this outpost the Western powers hoped to be able to roll up the growing people's democratic order when the occasion would arise. These plans were extraordinarily encouraged by the fact that in the wake of the elections of October 20, 1946, and as a result of the treacherous attitude of the right-wing Social Democratic leadership, which had prevented working-class unity in the whole of Berlin, the reactionary forces with the help of the Western occupation powers succeeded in gaining control of a great number of local government bodies which they turned into an instrument of their own policies and of the Anglo-American circles backing them, despite strong resistance by the democratic forces headed by the Socialist Unity Party.
>
> (Stulz 1961: 38)

It could even be held that

> in view of the anti-democratic character and the splitting aim of the terrorist elections of December 5th, these elections were not valid, and the West Berlin separate administration based upon the elections had no legal right to act as the representative of Berlin and its population.
>
> (Stulz 1961: 72)

In fact, the Social Democrats and their allies were to a considerable extent prevented from turning their electoral predominance into administrative performance by the activities of organizations for 'direct political action', such as the unified trade-union movement (FDGB), the 'Free German Youth' (FDJ) and the 'Democratic Women's League'. The Markgraf 'People's Police' remained excluded from the control of the *Magistrat*, even in the western sectors. In the last resort Social Democratic measures could be blocked by the SMA. Appeal to the Kommandatura was futile, as the unanimity rule gave the Soviet representative an effective veto on the decisions of that body. A major conflict emerged over the attempt to replace the SPD *Oberbürgermeister* Otto Ostrowski, who had resigned after being attacked for excessive complaisance towards the Soviet authorities, with Ernst Reuter, newly returned from exile in Turkey.

At the same time, events elsewhere were casting their shadow over Berlin. In the wider international sphere, relations between the western Allies and the Soviet Union took a turn for the worse in 1947, as the dividing lines in Europe hardened, and the Cold War set in. At the beginning of the year, the economic merger of the UK and US zones of occupation into the so-called Bi-zone began the process that was to end in the creation of the German Federal Republic. The inclusion of the western zones in the programme for Marshall Aid announced in June 1947 further emphasized their distinctiveness. It was in fact the increasing association of the western sectors of the city with the new economic organs in West Germany that provided the occasion of the final break. There had been intermittent interruptions to Berlin traffic, mainly to military trains, from March onwards, as an indication of the precarious nature of the Allied presence in Berlin, but the real crisis was signalled by the announcement on 18 June of a new currency for the western zones. Within a few days a separate currency reform was instituted for the Soviet zone, and it was unilaterally announced by the SMA that the new currency would also be adopted for the whole of Berlin. If accepted, this act would have brought the whole of Berlin under exclusive Soviet control. Initially, it was the elected Berlin Assembly that refused to accept the validity of this edict, not the western Allies, who vacillated in the hope of finding a compromise that would preserve Berlin unity. By 23 June, however, the western Allies had finally decided to allow the West Marks to circulate in the western sectors: this may be regarded as the decisive event determining that Berlin would henceforth be a divided city.

On the political front, there was a progressive division of administrative departments, beginning with the police and the food office. The regular meeting of the City Assembly on 26 August and the adjourned meeting on 27 August, held in the city hall in the Soviet sector, were broken into by Communist demonstrators, or as alternatively stated: 'August 26 and 27, Berlin workers demonstrate in front of the City Hall against the division of the city and for the normalization of the situation' (Stulz 1961). A further attempt to meet on 6 September was even more violently broken up by demonstrators who were openly directed by SED leaders. Assembly members other than those of the SED found the situation intolerable, and withdrew to meet in West Berlin. Henceforth there were to be two city administrations in Berlin. Alternatively described:

> September 6, 1948. Thousands of Berlin workers demonstrate again in front of the City Hall for a united Berlin. In reply, the reactionary majority in the city parliament illegally transfers its activities to the British sector and without consulting the representatives of the German Socialist Unity Party decides to hold new elections in late Autumn.
>
> (Stulz 1961)

In the Soviet sector, an alternative city government was installed, claiming to be the one fully legitimate body:

> On November 29th Ottomar Geschke, Deputy Chairman of the city parliament summoned an extraordinary session of the city parliament for the next day. Invited to the session were the city councillors and borough concillors of the whole of Berlin, and the parties and mass organizations that had come together in the 'Democratic Block'. Following a call issued by the FDGB more than 1,000 factory delegates for this extraordinary session were elected at about 800 factory meetings. Altogether about 1,600 legitimate representatives of the Berlin population came together in the Admiralspalast [a Berlin theatre] on November 30th. Acting on a resolution introduced by the Democratic Block they unanimously declared that the old City Council had been removed from office because it had 'ignored the most elementary interests of Berlin and its population and had continuously broken the constitution'. At the same time delegates decided to set up a new city council. On the proposal of the Democratic Block they unanimously elected Friedrich Ebert, Chairman of the Brandenburg Provincial Parliament, as Chief Burgomaster of Greater Berlin. Three new Deputy Burgomasters and a 15-man City Council were elected consisting of members of all the democratic parties together with the FDGB, the Democratic Women's League and the League of Culture. . . . The events of November 30th 1948 marked a turning point in developments in the German capital. The relation of forces in Berlin, and further in the whole of Germany, were changed to a decisive degree in favour of peace, democracy and progress. The constitution of the new Democratic City Council finally pushed the reactionaries back into the western sectors of Berlin.
>
> (Stulz 1961: 76–7)

The new 'Democratic City Council of Greater Berlin' was recognized as the only legal Berlin city administration by the acting Soviet commandant in Berlin on 2 December. Meanwhile, a somewhat more orthodoxly organized electoral process in West Berlin resulted on 5 December 1948 in a clear SPD victory, which was followed by the re-election of Ernst Reuter as *Oberbürgermeister*.

In the meantime, the Soviet authorities had first discovered various defects needing the indefinite closure for repairs of the road, rail, and water routes to Berlin, followed on 24 June 1948 by the explicit introduction of a total blockade. There were many among the military authorities in London and Washington who believed that the position of the western Allies in Berlin was untenable in the face of the blockade, and that withdrawal was inevitable. The alternative, the air lift, was almost the sole responsibility of one man, the US Commandant Lieutenant-General Lucius D. Clay. It must, however, be said that the prime decision having been taken by Clay, he received decisive political backing from President Truman in the United States and the UK and foreign secretary, Ernest Bevin, as well as quite decisive support on behalf of the Berlin population from Ernst Reuter. Within one day of Clay's decision on 24 June 1948 to institute the air lift, the US Air Force commander Lieutenant-General Curtis LeMay had the first food-laden C-47 landing at Tempelhof airfield. While the main effort was American, the United Kingdom also provided its quota of planes, including ageing flying-boats which landed on the Havel. The peak of the airlift was reached in April 1949, when a million tons of supplies were delivered; in one 24-hour period on April 15–16 a record 13,000 tons arrived. With 3,946 landings and take-offs, this involved an aircraft movement every 22 seconds. A plane aborting its landing was allowed no second chance, but had to return to base in West Germany to be allocated another space in the queue. To supplement Tempelhof and the British military airport at Gatow, a new hard runway was created in three months at Tegel in the French sector, by the labour of 20,00 Berliners. When the transmission tower of Radio Berlin, in the western sectors but in Soviet hands since 1945, proved to be in the way of aircraft movements, the French risked an international incident by simply blowing it up (Tegel is now the main West Berlin civil airport). One of the most astonishing achievements of the air lift was the flying-in of equipment for a complete electric power station, broken down into the smallest possible components (Anderhub, Bennett, and Reese 1984).

The cost (mainly to the Americans) of maintaining approximately 380 planes in use day in, day out was, of course, enormous, and accidents took a toll of seventy-eight American, British, and German lives. The airlift was a triumphant success, but only effective because of the wholehearted support, patience, and endurance of the Berlin population. The household coal ration for the whole of the bitter Berlin winter of 1948–9 was a mere 12.5 kg, which could be contained in a shopping bag. Electricity was available only for a few hours a day, often in the middle of the night, and gas supplies for cooking were strictly rationed. The food rations were delivered mainly in dried and powdered form, to save weight on the air lift. Although a few products were exported bearing the proud stamp of the Berlin bear breaking from his chains, it was impossible to bring in raw materials when even coal to meet minimum energy requirements had to be carried by air, so hundreds of firms had to close down, and unemployment soared; West Berlin was initially unable to partake in the 'German economic miracle'. Yet the West Berliners carried on, and few accepted the offer to register for more generous food supplies with the East Berlin authorities.

Faced with the success of the air lift, and conscious of the damage inflicted on the fragile economy of the Soviet zone by allied counter-measures, the Soviet Union was by early 1949 ready for an agreement to end the blockade; the barriers to land routes were lifted on 12 May. At the onset of the blockade, the Soviet Union was presumably anticipating an ignominious ejection of the western Allies from the city, which would have fallen entirely under their control and perhaps led to a collapse of the Western position in Germany as a whole. By the end of the blockade, the Soviet Union was clearly prepared to settle for the lesser objective of total control over the Soviet zone within a divided Germany, and at least *de facto* total control over East Berlin. It had, however, to be accepted that the blockade and air lift had produced an incredible sense of identity between the western Allies and their former enemies, and that West Berlin would henceforth be firmly linked with the emerging new state in West Germany, a position underlined by the decision of 21 March to make the West Mark the sole legal currency in that part of the city. Both sides, however, had to accept that any notion of a reunited Germany over which either would exercise significant influence was indefinitely deferred.

2.3 The constitutional situation

2.3.1 West Berlin

With the conclusion of the blockade, the western Allies gave West Berlin virtual political independence. In the initial version of the Basic Law (*Grundgesetz*) of the German Federal Republic, which was formally established on 21 September 1949, Berlin was included as a constituent federal state (*Land*) with the same rights as the other states, but this proposal was vetoed by the western Allies as being contrary to their conception of the whole of Berlin being under a peculiar Allied jurisdiction. Similarly the western Allies in 1952 vetoed the extension to Berlin of the law relating to the Federal Constitutional Court, stating:

> One of the principal functions of the Federal Constitutional Court is to resolve differences between the constituent parts and organs of the Federation. The adoption of the Federal Constitutional Court by Berlin would make it appear that Berlin was a constituent part of the Federation.

(Im Überblick: Berlin 1987)

In later years, with German reunification indefinitely postponed, the Allies were to become much less insistent that Berlin did not possess the status of a federal *Land*, and it was the Soviet Union that pursued the question (see section 2.6).

The constitution of West Berlin, as it came into force on 1 September 1950, maintained that West Berlin is a constituent *Land* of the Federal Republic, where the Basic Law of the Federal Republic is binding. Nevertheless, the peculiar status of West Berlin, stemming from the Allied occupation, requires that its representatives in the Bundestag (Lower House) and the Bundesrat (Upper House) of the federal parliament are not directly elected, but indirectly derived from the Berlin parliament (Abgeordnetenhaus). They have the right to speak but do not vote on legislation in its final form or in the election of the federal chancellor. However, Berlin representatives take a full part in discussion at committee stage and vote in the selection of the parliamentary Presidium. A Berliner can even become federal chancellor.

In their original acceptance of the Basic Law of the Federal Republic the Allies insisted that federal legislation would apply to Berlin only if voted as a Berlin law by the Berlin parliament. What happens in practice is that under the federal German 'Third Transfer Law' of 4 January 1952, Berlin is obliged to take over all federal legislation that is identified by a special 'Berlin clause' as applicable to the city. Effectively, this involves all federal legislation except items related to defence and emergency powers. Legislative uniformity between the Federal Republic and Berlin is thus maintained except in fields reserved by the occupying powers. Berlin also forms an integral part of the financial system of the Federal Republic. Through the intermediary of the Federal Republic Berlin also belongs to the European Community; its three representatives in the European parliament are indirectly elected by the Berlin parliament, but have full voting rights. Berlin's participation in this organization has been the subject of protests by the Soviet Union.

The special Occupation Status of West Berlin enables each of the western powers to maintain a garrison in the city, so that a Soviet or GDR military incursion would be a clear act of war. On the streets of the city, the most visible sign of the surviving Allied rights is provided by the freedom for Allied servicemen in uniform to circulate in all four sectors of the city. Apart from duty visits, leisure-time visits to East Berlin are popular with UK, US, and French servicemen and their families, groups of whom are often to be met strolling around. The application of Allied political rights in West Berlin is normally restricted to the following fields:

> The safety, interests and immunity from civil legal proceedings of the Allied forces
>
> the maintenance of a demilitarised status for the city
>
> the relationship of the city to outside authorities, including the Soviet Union
>
> ultimate control over the Berlin police
>
> *(Im Überblick: Berlin* 1987)

A somewhat gruesome survival of four-power joint administration of Berlin was Spandau gaol, where until 1987 the Nazi leader Rudolf Hess dragged out the last years of his life as sole prisoner. Following the death of Hess, the prison was razed to the ground to prevent it becoming a neo-Nazi shrine. A more useful survival is the Four-Power Air Traffic Control Centre. Another oddity is the presence just inside the UK sector on the 'Strasse des 17 Juni', the main ceremonial axis of Berlin, of the Soviet War Memorial. This has a ceremonial Soviet guard, the members of which are protected against molestation from the West Berlin civil population by a detachment of the West Berlin police. For security reasons, normal traffic is banned from stopping in the section of the 'Strasse des 17 Juni' in front of the memorial, but the inconvenience is minimal, as since the building of the Berlin Wall in 1961, the road has in any case been blocked for traffic at the Brandenburg Gate a short distance further east. For the most part, traffic consists only of tourist buses U-turning at the Brandenburg Gate to crawl past the memorial on their return. Military police of the western Allies can also be seen making routine patrols along the boundaries of the city, or putting up notices forbidding the holding of meetings and the use of noise-generating equipment in the vicinity of the 'Strasse des 17 Juni' on the rare occasion of a military parade. In the western sectors, Allied rights are also used to veto the presence of federal German armed forces and to excuse young men from the obligation to undertake military service, a decision that provides West

Berlin with the dubious benefit of a population inflow from the Federal Republic of men seeking to avoid this obligation.

Under its 1950 constitution, West Berlin is a state of the Federal Republic (subject to the restrictions stemming from Allied rights) and a city. Its administration accordingly combines functions that in the 'mainland' Federal Republic exist at two different levels (except in the other 'city states' of Hamburg and Bremen). The democratically elected Berlin parliament has nominally 200 members, but only 144 seats are occupied (1987), the remainder being reserved for the absent representatives of East Berlin. The parliament passes legislation relating to Berlin and elects the Berlin senate (government). From the division of Berlin to 1981 the SPD, with varying coalition partners, controlled the government. Power then passed to the Christian Democratic Union (CDU) under Richard von Weizsäcker, subsequently federal president; from 1983 the CDU was in coalition with the small group of Liberal Democratic Party (FDP) representatives. The Communists, from 1969 known as the SEW (the West Berlin equivalent of the GDR's SED) have never achieved more than 2.7 per cent of the votes. A feature since 1969 has been the rise of the unconformist 'Alternative List' (fifteen seats in the 1985 election).

The senate (state government) of West Berlin consists of the *Regierender Bürgermeister* (governing mayor), the *Bürgermeister* as his deputy, and not more than sixteen senators. The governing mayor and the other members of the senate, whom he nominates, have to achieve the support of a majority of members of the parliament. The governing mayor determines the main lines of policy in agreement with the senate, the members of which head the departments of the city administration, equivalent to ministries elsewhere. In addition, each of the twelve West Berlin *Bezirke* has its own elected district mayor (*Bezirksbürgermeister*) and assembly (*Bezirksrat*). The *Bezirke* have no powers to pass by-laws but carry out tasks that the city decides are better located closer to the people than at central level. They control schools, for example, and have their own building programmes (*Im Überblick: Berlin* 1987).

Berlin is also linked to the Federal Republic by flows of financial assistance. The West Berlin that emerged from the blockade was an economically crippled city. Its industry was cut off from suppliers and markets in the immediate hinterland, the territory that was in process of becoming the GDR. The Berlin internal market had also greatly diminished, and the city's new isolation put its manufacturers at a cost disadvantage as compared with those in West Germany. It is not surprising that what remained of Berlin industry after wartime damage and Soviet dismantling was only operational to 40 per cent of capacity. Unemployment in 1950 reached a total of 289,000 persons, or

37 per cent of the workforce, and all this at a time when the economy of the Federal Republic was soaring into the 'German economic miracle'.

The economic recovery of West Berlin began with its inclusion in the Marshall Plan in 1949, supplemented by assistance from the federal government that has continued to the present day in successive versions of the *Berlin-Hilfe-Gesetz* (now *Berlinförderungsgesetz*). The general objective of economic assistance is that living conditions in West Berlin shall not fall below those of the Federal Republic. The nature of assistance to the economy has varied somewhat over the years, as economic conditions have changed, but in general three broad categories can be distinguished. The first category consists of various tax preferences; for example, a reduced rate of turnover tax assists the sale of West Berlin goods and services in the Federal Republic, while favourable rates of write-off and various tax allowances are designed to stimulate investment. Other tax concessions are designed to make West Berlin financially attractive for employees; those moving to Berlin also receive special transitional assistance. These forms of assistance came particularly to the fore after 1961 when, in a period of full employment, it was difficult for Berlin to attract and hold workers.

A second category of measures embraces various forms of federal subsidies and special credits to assist the development of industry, craft concerns, and even agriculture and horticulture, as well as for such special purposes as the erection of buildings of cultural significance or the care of non-German refugees. The federal government has also done something to compensate West Berlin for the loss of income and employment relating to capital functions by locating there some of the less sensitive government offices (see section 5.4.4).

A third and particularly important form of federal help relates to transport to and from West Berlin, reducing the impact of the city's isolation. The federal government in fact insures goods moving to and from Berlin against politically motivated interruption. It subsidizes the movement of some heavy commodities and provides telephone and postal rates below those that would normally be charged for the distance involved. Air tickets between West Berlin and the Federal Republic are also subsidized. After the 'normalization' of relations associated with the 1971 Berlin Agreement, the Federal Republic was to make a number of direct deals with the GDR to improve access to Berlin, including the buying-out of various GDR exactions on transit traffic and payments for the improvement of autobahn and rail access to the city.

In addition, what would otherwise be a large budgetary gap between what the city spends on the support of its population and what it receives from them in taxes is made up under a scheme for direct financial assistance from the federal government. The sum involved has risen steadily, both in absolute terms and relatively; in the 1960s it was about 40 per cent of the Berlin budget, reaching 54 per cent in 1981 (1986, 52 per cent). The sum agreed by the federal government for 1987 was DM 12 billion (i.e. thousand million) out of a total Berlin budget of about DM 23 billion. These various forms of assistance, involving some enormous sums, are the price paid by the Federal Republic for the maintenance of this 'shop window of the west'.

2.3.2 East Berlin

The armed forces of the Soviet Union, like those of the western Allies, have the right to circulate throughout the whole of Greater Berlin. The right to station troops in the city carries, however, none of the significance that it has for the western Allies, since the Soviet Union has massive military forces based on the territory of the GDR, which could be called upon in any emergency, as they were during the unrest of June 1953.

At the somewhat superficial level of the rights of military personnel and diplomats of the occupying powers to pass unhindered between the two parts of the city, the special four-power rights in Berlin are still accepted by the Soviet Union and, with manifest reluctance, the GDR. It was made clear in the course of the negotiations for the Berlin Agreements of 1971 that the Soviet Union still regards itself as having some special responsibilities for the city. It is however the position of the GDR that East Berlin is an integral part of its territory and, indeed, its capital, a role underscored by the clumsy official title 'Berlin, Hauptstadt der DDR' (Berlin, Capital of the GDR).

The vast majority of the restrictions inherent in the four-power status of Berlin have been unilaterally removed so far as East Berlin is concerned, particularly those requiring the demilitarization of the city. Since 1956 units of the GDR National People's Army and of the armed workers' militia have regularly paraded in East Berlin on May Day and other festivals, to the unavailing protest of the western powers. The Berlin Wall of 1961 was built under armed protection, and remains so guarded to this day. Once the Wall had been completed, the way was clear in 1962 for the extension of compulsory military service to East Berlin. In the same year the Soviet Union abolished its Berlin Kommandatura (garrison headquarters), which was replaced by the 'City Commandant for the Capital of the GDR, Berlin'. From 1981 the inhabitants of East Berlin were also allowed to participate in the direct election of representatives of the GDR *Volkskammer* (the parliament).

Constitutionally, Berlin is equivalent to a *Bezirk* of the German Democratic Republic, with its own elected assembly and administration (*Magistrat*) centred upon the 'Red Town Hall' in the centre of the city.

There are also subsidiary administrations for the various city districts, as in West Berlin.

Superficially, the East Berlin administration would seem to have more responsibilities than its counterpart in West Berlin, because state control extends over a much wider range of human activities than in the west. Reality is rather different, because of the highly centralized nature of the state in the GDR. The role of the East Berlin administration, as in all of the GDR *Bezirke*, is to implement policies laid down by the government. The officials who implement policies at this level tend to have a career pattern that looks to advancement within the structure of national ministries, which may modify their willingness to devote themselves wholeheartedly to fighting for the satisfaction of local needs (Childs 1983). A number of functions that might elsewhere be carried out by local government are also in the GDR in the hands of separate nationally based organizations. Tourism in Berlin, for example, is essentially controlled by the 'Berolina' national tourist organization, while higher education comes under the GDR Ministry for Universities. Bureaucratic centralization can have advantages for Berlin, which is something of a 'show-case for socialism'. For example, in the run-up to the 750th anniversary celebrations, the other *Bezirke* of the GDR were required by the government to send teams of builders to work in Berlin, much to the disgust of the local populations.

The other restriction on the powers of Berlin local government is provided by the Communist Party. Every institution in the GDR has a parallel SED organization, which transmits to it party policies, and controls that they have been properly implemented. So just as the first secretary of the SED is the most powerful person in the GDR, so the first secretary of the Berlin SED, not any elected figure, is the most powerful person in East Berlin. If the powers of elected representatives thus appear very restricted, it can be argued that democracy is extended by the voluntary participation of Berlin residents in a whole range of committees and activities, naturally guided and stimulated by the party, ranging from leadership groups in residential complexes to auxiliaries of the People's Police. In 1985, 163,600 such positions were occupied by East Berliners.

2.4 Berlin: shop window of the west

A major milestone was passed in 1957, when West Berlin's overall production for the first time exceeded its internal consumption, thus leaving a surplus derived from sales elsewhere. Following the end of the blockade, Berlin's economy began to blossom, and as it did so, 'so did the prosperity of its commercial establishments. Stores again were well stocked; hotels and restaurants began to thrive; neon lights went on at night, and the *Kurfürstendamm* and *Tauentzien* once more became alive with shoppers and tourists' (Smith 1963: 140).

Even during the period of the blockade, the inhabitants of Berlin had been able to pass freely from one part of the city to another. This curious situation of political division without total physical separation continued until 1961. The boundary between East and West Berlin was clearly marked, but could be crossed without a permit; GDR police controls were the exception rather than the rule. For example, at the Potsdamer Platz, once the 'Piccadilly Circus' of Berlin and now in a wasteland along the sector boundary, the East Berlin tram halted and it was only necessary to walk across to pick up the tram or bus on the West Berlin side. Those unwilling to risk the possibility of being checked by the East Berlin police could take the S-Bahn or U-Bahn, which continued to circulate throughout the entire city. A whole way of life developed along the sector boundary. Hastily repaired shops, wooden shanties, and even market-type open stalls appeared at the most frequented crossing points, such as the Potsdamer Platz or Spandau-West S-Bahn station. They offered scarce foodstuffs and the products of western consumerism at rock-bottom prices to such inhabitants of the GDR as had western currency, or were willing to buy it at a heavily unfavourable rate in the exchange booths that also sprang up. The same 'black' rate of exchange made it possible for West Berliners to drain East Berlin of the admittedly rather limited range of commodities available there. Supermarkets were even set up on the West Berlin side deliberately angled at East Berlin purchasers, for example at Gesundbrunnen, the first S-Bahn station west of the sector boundary to the north of the centre.

The federal government and the West Berlin senate also subsidized visits for cultural purposes by people from East Berlin and the GDR; they could pay in GDR currency for visits to theatre, opera, and concert performances, or to see films decreed to be of cultural importance. In the 1950s, one-fifth of visitors to West Berlin museums came from the GDR and East Berlin, and about 30 per cent of students in universities and colleges. Visitors could also use GDR money on the public transport system. They could even fly from Tempelhof to visit the Federal Republic (Engert 1985). It is not surprising that from the sober austerity of East Berlin, where central-area reconstruction took off much later, the authorities looked dubiously at the dazzle and at least superficial prosperity of the new West Berlin, as well as at its cultural attractions. There were other problems. The existence of the open sector boundary meant that the GDR and its occupier did not entirely abandon the practice of kidnapping political opponents that had been normal before the blockade in the time of the Markgraf police. On the other hand, the GDR authorities had right on their side in condemning West Berlin as a nest of espionage agents, taking advantage of a privileged situation well behind the Warsaw Pact front line, but espionage is a game at which two can play. Although espionage activities in Berlin must have been reduced since the building of the Wall in 1961, the city has remained a favourite *mise en scène* of spy novels (Deighton 1964).

Berlin's location closer to Poland than to the Federal Republic also meant that the broadcasts from SFB, the West Berlin radio, and from the extremely powerful United States-backed RIAS, could be heard throughout the GDR. Subsequently, the same was to be substantially true of television; only a few remote pockets could not pick up transmissions either directly from the Federal Republic or from SFB. The GDR was and is thus unique among the Soviet satellites in not having total control of the information conveyed to its citizens. The greatest problem was, however, that the open sector boundary enabled any person disenchanted with life in the GDR to leave for the west, even after the direct boundary with the Federal Republic had been sealed. Such persons knew that, after a fortnight in emergency accommodation in West Berlin, they would be flown out to the Federal Republic, and hundreds of thousands of people took advantage of the opportunity. From the GDR point of view, the position was made worse by the fact that those who left consisted to a high degree of the younger, most active, and most skilled elements of the population, some of whom had been trained for their profession at great public expense.

2.5 Krushchev and the Berlin Wall

In view of all these difficulties, it is hardly surprising that the existence of the island of West Berlin was regarded with abhorrence by the authorities of the GDR, and that the period between the end of the blockade to the present day has been marked by efforts by the GDR, with qualified support from the Soviet Union, to neutralize and, preferably, to remove it.

In the continuing dispute over West Berlin, the western powers on the one hand and the Soviet Union and the GDR authorities on the other maintain two irreconcilable points of view, which are set out below in a form drawing heavily upon the work of Wettig (1981, 1986). The three western powers take the view that their original occupation of Berlin alongside the Soviet Union gives them equal rights with the Soviet Union over the entire city. Being in Berlin as of right, it follows that they also have a right of access to the city. While it is accepted that the western powers currently operate a three-power occupational rule restricted to the western sectors, with only some minor military survivals covering the city as a whole, it is claimed that this is merely the inevitable consequence of the withdrawal of the Soviet Union from the Inter-Allied Kommandatura, and does not involve any reduction of the claims of the western Allies in relation to the city as a whole. Given the four-power nature of the Berlin occupation, which has not been terminated by the conclusion of a formal peace treaty, it follows that the Soviet Union cannot unilaterally transfer its powers over the Soviet sector to the government of the GDR. The western powers also hold that, since the creation of the GDR was subsequent to the establishment of the occupation arrangements for Berlin, it is incumbent on the GDR authorities to respect these prior arrangements, which remain valid.

The point of view of the Soviet Union, asserted with particular vehemence during the 1948–9 blockade, and consistently since 1958, is that Berlin since 1945 was never other than an integral part of the Soviet zone. It is admitted that in 1945 the Soviet Union permitted the western Allies to maintain a presence in the western sectors, but this was conditional on the acknowledgement of overriding Soviet authority. Wettig's view that this assertion has no basis in the fundamental Allied agreements on Berlin reached in 1944–5 appears to be confirmed by the extracts from documents reproduced above (section 2.1), but 'Moscow acts on the thesis as if it were a well established and universally acknowledged fact' (Wettig 1986: 87). It follows from this fundamental Soviet position that the claim of an original western right of presence in Berlin is rejected in principle; it is only by 'grace and favour' of the Soviet Union that a western presence in Berlin is tolerated. Even this tolerated occupation was intended to be a purely transitional arrangement, certainly not to last for forty years or more, so that the western Allies should have packed up and gone years ago. There are also, in the Soviet view, no western access rights. Responsibility for military transit rests with the Soviet Union, while responsibility for civil transit has been transferred to the GDR authorities.

It is further asserted that the Soviet Union, as exclusive holder of full occupational authority in Berlin, has transferred its rights regarding the Soviet sector to the GDR. For this reason, quadripartite authority, including the authority of the Soviet Union, pertains only to the western sectors; there is no parallel quadripartite authority over East Berlin, as the western powers claim. It follows that any decisions regarding West Berlin which have not been taken with full quadripartite authority, including the agreement of the Soviet Union, are invalid, which virtually means that all decisions since 1946 have been invalid. This applies particularly to the links that have been forged between

West Berlin and the Federal Republic. West Berlin, so long as it continues to exist, is a political entity of a special sort, entirely dependent in its constitution on the agreement of the Soviet Union. Admittedly, as will be seen, certain matters relating to West Berlin were eventually regulated by the Berlin Agreement of 1971, but any points not embraced by the 1971 Agreement must still be regarded as open until such time as Soviet consent is obtained. It further follows that since the western powers have no rights except by favour of the Soviet Union, they have no rights in relation to its successor, the GDR; this particularly applies to access to West Berlin across GDR territory.

There were certainly moments after the end of the blockade when it seemed possible that Berlin's history might have gone another way. Stalin, in March and April 1952, appeared to be offering to sacrifice the GDR in favour of a united but neutral Germany, within which Berlin would resume its role as capital. It is still uncertain whether Stalin's offer was genuine, or whether he was bluffing, merely hoping temporarily to disrupt the political process leading to the final termination of the occupation status in the Federal Republic and the integration of the Federal German army in a western defence community (Bucerius 1985). Similarly, vague proposals for a unified and neutral Germany reappeared after Stalin's death, but on the western side only Ernst Reuter, governing mayor of West Berlin, appears to have regarded them as a serious basis for negotiation. The United States was embedded in the Cold War, fearful that a neutralized Germany would involve the total withdrawal of its military forces from mainland Europe, leaving Soviet forces poised on the Oder, an hour or two from Berlin and a day or two from the Rhine. France, of course, had never been a partisan of German unity. The ageing Adenauer was equally unenthusiastic, conditioned by memories of past political battles with Protestant Prussian Junkers and anti-clerical Social Democrats from Berlin. He was much more drawn towards his co-religionists in Western and Southern Europe, the participants in the future EEC. As much as anything, Adenauer's establishment in 1955 of diplomatic relations between the Federal Republic and the Soviet Union signalled the death of notions of German unity by agreement (Francisco 1986).

The 1956 Hungarian uprising, following the 1953 near-rebellion in the GDR, revealed to the Soviet Union not only the need to strengthen its control of its satellites, but that the West was no longer in the business of rushing into Eastern Europe as a liberator. To Krushchev, it may well have seemed that the West would at this point regard Berlin as expendable, no longer needed as a window upon an Eastern Europe that was relegated to the Soviet sphere of influence. Hints at the possibility of a united if neutral Germany having failed to halt the consolidation of a rearmed Federal Republic into the western camp, Krushchev evidently decided to see what open pressure would accomplish. On 27 November 1958 the Soviet ambassadors in Washington, London, Paris, and Bonn delivered a lengthy and hectoring note, stating that repeated western violations of the Potsdam Agreement meant that the agreements reached in 1944–5 allowing the participation of the western Allies in the administration of the city were quite simply null and void, and the presence of the western Allies therefore illegal. The western Allies were required to leave, after which West Berlin was to become a free city, in which neither the Federal Republic nor the GDR would interfere, and all this was to happen within six months. The ultimatum was backed by a threat to transfer all control over access to Berlin to the GDR, without Soviet intervention to protect western rights. By January 1959 this had become a more specific threat to conclude a separate peace treaty with the GDR. In the background there was also the possibility that warlike measures would be taken, should the western Allies insist on remaining in Berlin (Windsor 1963: 200–2).

Almost unbelievably, after long negotiations at the highest level, nothing happened; the six-month time limit was first extended, then forgotten. The West gained time, and Krushchev, faced with the American nuclear arsenal, backed away from any military action. Perhaps the only gainer was the GDR leader Ulbricht, who in the course of negotiations had firmly established his state as a partner to be involved in any future arrangements regarding Germany and Berlin. The next decisive move was definitely Ulbricht's, the extraction of Soviet agreement to the sealing-off of West Berlin.

By 1960 the GDR seemed to be approaching collapse. Agriculture was in chaos, following enforced collectivization, and industrial production was faltering. In the summer of 1961, refugees were entering West Berlin at a rate of over 1,000 a day. Krushchev appears to have concluded from his June 1961 meeting with President Kennedy that the essential American concern was with the freedom of the West Berliners to determine their own form of government and with freedom of access. There was at least the implication that the western powers accepted that something would need to be done to stabilize conditions in the GDR, and that the West would and could do nothing in response (Francisco 1986: 12–13). Ulbricht was accordingly able to extract Krushchev's agreement to the building of the Berlin Wall.

On 12 August 1961, in anticipation of new measures to restrict movement, over 4,000 refugees crossed into West Berlin in a little more than twenty-four hours. In the early hours of Sunday 13 August armed factory-militias and police, backed by units of the GDR army, sealed off the sector boundary between West and East Berlin.

Immediately, work began to tear up the roads and to erect an initial barbed-wire barrier to divide the city. Civilian movement between the two parts of the city was virtually stopped; not only was the flow of refugees dammed overnight, but the *Grenzgänger*, who had lived in East Berlin but had worked in West Berlin, were obliged to find new jobs. They had at one time been officially numbered at 56,300, to which were added an estimated 20,000 persons, mostly female, undertaking domestic work or engaged in a second occupation. The West Berlin building industry was particularly hard hit, while the loss of some 8,000 women workers dealt the West Berlin clothing industry a blow from which some firms did not recover. Since West Berlin residents working in East Berlin had numbered only 12,000 (and many of these in fields like education or the arts), the interruption of this exchange provided a valuable boost to the East Berlin labour force.

At the same time the S-Bahn and U-Bahn rail systems were separated; West Berlin trains continued to use three tunnels beneath East Berlin, but they rolled without stopping past darkened and guarded stations. The only East German station to which U-Bahn and S-Bahn travellers from the West continued to have access was at the heavily guarded Friedrichstrasse crossing-point. Although thirteen crossing-points were initially left open (the number was later much reduced), the action was clearly incompatible with western views of a joint four-power occupation regime covering the whole city. Nevertheless, the West stood by and did nothing; East Berlin became effectively part of the German Democratic Republic, over which the western Allies had no powers other than an increasingly pointless right of entry for their uniformed military personnel.

The initial temporary boundary was progressively transformed into the 'Berlin Wall' that we know today. It is far more than the simple, graffiti-covered wall that turns its face to the West; behind it lies a whole defence in depth, with alarm wires, raked death-strip, prowling dogs, floodlights, and watch towers. All this is not just for display; if escapers are extremely few, from time to time one more unfortunate falls beneath the bullets of the frontier guards. The Wall follows not only the sector division but the common boundary of the GDR with West Berlin, which is thus totally encircled. Crossing from West to East Berlin involves, for all but the Western armed services and diplomats, the same formalities as crossing any international frontier, if not rather more so. A person entering (and even more particularly one leaving) must stand in a sort of steel box, to be interrogated through a security window by an officer of the frontier police, and only allowed to proceed when passport and visa details have been checked against central records.

The GDR official explanation of the Wall is that it is a defence

Plate 2.1 The western face of the Berlin Wall is a gift for graffiti writers. Beyond the Wall, the tramlines can just be discerned running through the former Leipziger Platz, once at the busy heart of pre-1945 Berlin, now part of the Wall's death strip. Beyond are East Berlin government buildings along the Otto-Grotewohl-Strasse (formerly Wilhelmstrasse) and in the far distance new residential developments along the Leipziger Strasse.

against external attack from the 'aggressive imperialist forces in the FRG and in other NATO countries'. In fact, by its creation the threat of internal collapse had been brutally but effectively eliminated. With the possibility of flight removed, the traditional German virtues of industry and obedience re-emerged. The GDR became perhaps the most tightly planned, tightly controlled of all the Soviet-style states of Europe. The economy was so far transformed that even western observers spoke of an 'economic miracle' in the 1960s that rivalled West Germany's astonishing growth in the 1950s. The western powers presumably thought that acceptance of the Wall had served to contain a deteriorating situation at an acceptable level, without sacrificing West Berlin's essential liberties, but the people of West Berlin were left isolated and insecure, shocked by the inability or unwillingness of the west to do anything to prevent the savage division of their city.

2.6 Détente

After the construction of the Wall, the Soviet Union continued for a year or two with the familiar pattern of minor harassment, particularly along the lines of communication, designed to chip away at the western position in Berlin. By the mid-1960s, however, the major powers had enough preoccupations elsewhere to be willing to consider an agreement that would retire Berlin from the forefront of international political attention, and make it unlikely that the city would become a flashpoint for global conflict. At a more immediate level, the western powers wished to improve the position of West Berlin, freeing it from the repeated Soviet-induced or GDR-induced crises, particularly with regard to access, which tended to produce feelings of insecurity in the population and which were economically damaging. A major objective was to reduce West Berlin's isolation.

It might be thought that, at local level, the Soviet Union had every incentive to keep the Berlin pot on the boil; this was certainly the view of the Ulbricht regime. The Soviet Union, however, was at this time concerned to stabilize the political system in Central and Eastern Europe, and it also needed to obtain vitally needed economic help for itself and the GDR. To do so it needed the completion of treaties with the Federal Republic, which made it clear that no progress would be made without a satisfactory agreement on Berlin. From the eastern point of view, negotiations on Berlin were a concession to achieve other aims. This did not mean that the Soviet Union had no specific objectives of its own with regard to Berlin. It was determined that, whatever the form of words used, any agreement should in practice apply only to West Berlin, and should in no way restrict the existing rights of East Berlin. It was also determined to do everything possible to counter one of the principal western objectives, which was to reduce West Berlin's isolation through the strengthening of ties between West Berlin and the Federal Republic.

Although the 1971 Quadripartite Agreement on Berlin was in many ways only a recognition of the situation that had existed in Berlin since 1961, it had from the western point of view the advantage that it involved a degree of recognition by all four of the occupying powers, and this alone went a long way to eliminate the uncertainty and stress previously experienced by the West Berlin population. It greatly eased restrictions on travel to and from West Berlin and reopened access for West Berliners to the GDR and East Berlin. Following the agreements, it has also been generally accepted that West Berlin should be represented externally by the Federal Republic, but with some definite limitations. In particular, the Soviet Union has blocked the adequate inclusion of West Berlin in its various agreements on co-operation with the Federal Republic and steadfastly campaigned against the inclusion of West Berlin members in official Federal German delegations to international organizations and international conferences.

If there were some real gains, there was a price to be paid. The agreement left West Berlin formally under the political control of the western Allies of the Second World War, a curious arrangement twenty-six years after the close of the war, and even more curious today (Francisco 1986: 19). It specifically stated that West Berlin is not 'a constituent part of the Federal Republic of Germany, and not to be governed by it'. While not all forms of association with the Federal Republic were forbidden, officials of the Federal Republic were proscribed from carrying out certain duties in West Berlin, and the Soviet Union was given a voice in deciding what was permitted, and what was not. Furthermore, while the western Allies formally regarded the Berlin Agreement as applying equally to East Berlin, in fact the agreement was not symmetrical, being operational only in West Berlin. The GDR's claim that East Berlin is an integral part of its territory and its capital was tacitly accepted, a fact underlined by the subsequent establishment of embassies of the United Kingdom, the United States, and France and a 'Permanent Representation' of the Federal Republic there. The concept of a 'Greater Berlin' was abandoned, the possibility of a unified Berlin postponed indefinitely, with a marked impact on the way that West Berliners perceived the future development of their part of the city.

The agreement was in any event only a *modus vivendi*. At no time did the Soviet Union or the GDR abandon their position that West Berlin was an integral part of the GDR that happened to have a peculiar political regime, existing on a sufferance that could at any time be withdrawn. The Soviet Union did not totally restrain the GDR from minor interference with the access routes to Berlin, particularly on the occasion of events in West Berlin that suggested a connection with the Federal Republic. In 1978, for example, a visit of President Carter to West Berlin accompanied by Federal Chancellor Helmut Schmidt was marked by delays on the autobahn.

This is not to say that the GDR was happy with the agreement. Ulbricht would certainly have preferred no agreement at all, leaving West Berlin open to continuing incroachments. Before the Berlin Agreement was finally signed, the Soviet Union was obliged in May 1971 to install the somewhat more malleable figure of Erich Honecker as GDR leader (Francisco 1986: 20). The GDR was still faced with the continuing existence of West Berlin, perceived as an obstacle to the integrity of its territory and as a source of subversion.

If Soviet policy at the moment excludes a forcible resolution of the Berlin problem, clearly the GDR sees no reason why it should be obliged unduly to strive to keep West Berlin alive and thriving. It has, for example, consistently and, on the whole, successfully opposed the notion that West Berlin might be built up as an economic and cultural crossroads linking east and west, perhaps with the location in the city of agencies of the United Nations, which West Berlin would dearly like. In this view, West Berlin's status is certainly not to be enhanced; rather it is hoped that the city will wither on the vine, falling eventually into the lap of the GDR.

The GDR also takes any opportunity that presents itself to nibble away at the status of West Berlin, of recent years focusing particular attention upon the nature of the boundary separating the two parts of the city. The western position is that this is a purely internal political division within Greater Berlin, the GDR position is that it is an international boundary. Following the logic of their position, the *Land* of West Berlin and the federal authorities make no attempt to operate immigration controls along this line. To put this position under pressure, in 1985–6 the GDR allowed thousands of people from low-income Asian and African countries to cross its territory to flood into West Berlin, taking advantage of the generous provision there for political asylum (see section 8.3.2). In this way the GDR hoped to oblige West Berlin to institute immigration controls.

On its own side of the Wall, of course, the GDR can institute whatever controls it likes, and in fact, both for emigration and immigration, they must be among the most rigorous in the world. Even here, however, the remaining four-power rights mean that unrestricted entry must be given to Allied military personnel in uniform, and similar privileges are claimed by diplomats of the western Allies accredited in East Berlin. In May 1986 the GDR authorities tried to

Plate 2.2 Berlin's main axis, the Unter den Linden, is barred-off on the East Berlin side, short of the Brandenburg Gate. The whitewashed face of the Berlin Wall can be discerned through the gate. At this point in 1987 young East Berliners, trying to hear a pop concert in West Berlin, clashed with the GDR police.

demand that such diplomats show their passports when crossing between East and West Berlin, instead of the customary diplomatic identity card. This demand was rejected by the western powers because it would have reinforced the GDR claim that the Wall is an international boundary in law. Western reaction was so vehement that the demand was withdrawn after a few days, but nobody would expect such attempts to nibble away Western rights to cease.

Relations between West and East Berlin have also been affected by the extraordinary degree of economic dependence of the GDR upon the Federal Republic that has grown up since the 1971 agreement. The GDR benefits from open access to the markets of the Federal Republic and, through the intermediary of the Federal Republic, to the markets of the EEC ('honorary membership of the EEC'). The GDR also depends on substantial financial credits from the Federal Republic, which can be used for such purposes as the purchase of equipment to modernize its industry. The leverage that the Federal Republic thus obtains has not been used to press for major political changes but rather for concessions on humanitarian grounds, for example with regard to travel or the reuniting of families. Some of these concessions, particularly those regarding travel, are of direct benefit to the West Berlin population. An exceptional rather more 'political' use of the Federal Republic's leverage came in 1986, when the GDR was persuaded to require that persons transiting its territory to enter West Berlin should have a valid West German visa, thus cutting off the embarrassing flow of 'political refugees'. In addition, the federal government buys out for cash various GDR impositions upon the unfortunate surface traveller to West Berlin, such as transit visas and special taxes on vehicles for the upkeep of the autobahn. It has also contributed massively to the cost of building a new Berlin–Hamburg autobahn, of rebuilding the Helmstedt–Berlin autobahn, and of reconstructing the Helmstedt–Berlin railway.

Détente has increased the number of issues relating to West Berlin on which negotiations with the GDR become necessary. In such cases it has become standard form for the GDR government to attempt to institute direct negotiations with West Berlin, as if the latter were an independent 'free city', unconnected with the Federal Republic. Such attempts are routinely opposed by the Federal Republic, which regards itself as alone responsible under the 1971 Berlin Agreement for negotiations between West Berlin and external authorities, including the GDR. The western Allies also oppose direct negotiations when they feel that their own powers are involved. Negotiations over the transfer of the West Berlin section of the S-Bahn to West Berlin were, for example, excessively protracted through GDR attempts to negotiate at government level with West Berlin, a difficulty only resolved when the

East Berlin-based administration of the *Deutsche Reichsbahn* was named as negotiating partner. Since most of these negotiations eventually involve the payment of large sums in scarce hard currency by the Federal Republic, the latter is usually in a strong position to ensure that the negotiating arrangements do not offend its position concerning its rights with regard to West Berlin. The matter came up in a peculiar form on the occasion of the 750th anniversary celebrations in 1987, when the West Berlin governing mayor, Eberhard Diepgen, invited the East German leader Erich Honecker to visit West Berlin. There was obvious relief on the part of the Federal Republic and the western Allies when Honecker refused the invitation, presumably on the grounds that he would have been placed at the opening ceremony alongside the president and chancellor of the Federal Republic, whose rights in West Berlin the GDR consistently opposes.

2.7 Living in a divided city

After the turbulent years between 1945 and 1961, the building of the Wall and the 1971 Berlin Agreement have brought to Berlin a period of relative tranquillity, signalled by the city's extremely infrequent appearances in the headlines of the international press. The fact that these arrangements, however precarious, have allowed some sort of peace to continue for decades encourages the hope that they will continue to do so.

Yet a great city has been divided, and West Berlin is so thoroughly sealed off both from East Berlin and from its hinterland that it is easier to list the few fields where some interconnection survives, than those where interconnection no longer functions. The extraordinary island city of West Berlin has all its most essential economic, political and cultural links not with its immediate environment but with the 'mainland' of the Federal Republic, 175 km and more away.

It has been a long-standing principle of GDR policy that West Berlin shall be prevented from building itself up to be an intermediary between East and West in Europe. Nevertheless, the GDR has had to accept a West Berlin location for the *Treuhandstelle für Industrie und Handel* (TSI), the office set up by the government of the Federal Republic as the negotiating partner with the section of GDR Trade Ministry dealing with the Federal Republic and West Berlin. Through the intermediary of TSI (until 1981 known as the *Treuhandstelle für den Interzonenhandel*) it was possible for the Federal Republic and the GDR to make trading arrangements even before the 1972 *Grundlagenvertrag* established the Permanent Representatives in Bonn and East Berlin. TSI officials meet with their GDR opposite numbers fortnightly, in East and West Berlin alternately, and also in Leipzig at the time of the Trade Fair.

Although the presence of TSI adds something to the importance of West Berlin, most of the trade with which it is concerned is between the GDR and the Federal Republic. Only about 1 per cent of all goods exported from West Berlin goes to the GDR and East Berlin, and about 1 per cent to other Soviet-bloc countries; this contrasts with about 77 per cent of West Berlin goods that go to the Federal Republic and 21 per cent for export. Imports from the GDR are slightly higher, but still small, running at about 7 per cent, compared with 73 per cent of requirements obtained from the Federal Republic and 20 per cent by imports (Schütz 1985) (Tables 2.1 and 2.2).

Table 2.1 *Deliveries from West Berlin to the GDR and East Berlin, 1984*

Commodity	Tonnes	DM (000s)
Iron and steel	110,837	78,753
Non-ferrous metals and semi-finished products	563	13,011
Machinery	521	35,051
Road vehicles	766	11,979
Electrical and electronic equipment	1,776	40,904
Fabricated metals	514	12,578
Pharmaceuticals	188	19,241
Other chemical products	1,213	10,302
Printing and reprographic products	799	14,249
Food and drink	4,257	10,065
TOTAL	135,611	315,994

The main West Berlin purchases from the GDR consist of heavy goods, where minimizing transport cost is an important consideration. Typical in this respect are building materials, which make up the largest category received when measured by weight. The GDR provides virtually all the sand and gravel used in West Berlin, as well as a substantial proportion of cement, stone, and paving materials. Nearly half of this material moves into West Berlin by water, with building rubble for dumping providing a useful return cargo. Second in terms of weight but first in terms of value is the group of refined petroleum products, including about 75 per cent of the diesel or heating oil and 70 per cent of the petrol consumed. These commodities arrive almost exclusively by rail. Energy supplies also include

Table 2.2 *Deliveries from the GDR and East Berlin to West Berlin, 1984*

Commodity	Tonnes	DM (000s)
Agriculture:		
vegetable products	133,948	75,101
animal products	47,096	128,158
Mineral products	571,425	52,621
Refined petroleum	1,535,521	1,198,671
Stone, sand, gravel, etc.	3,397,769	68,892
Iron and steel	90,627	37,544
Non-ferrous metals and		
semi-finished products	17,146	73,340
Glass and glass products	46,004	21,588
Clothing	4,787	160,356
Food and drink	79,401	99,171
TOTAL	6,093,978	2,092,094

virtually all the brown-coal briquettes consumed in West Berlin. It is, of course, a measure of the GDR's urgent need to earn hard currency that a country where energy supplies are constantly 'tight' should be willing to sell these commodities to West Berlin. Another important group consists of agricultural commodities, with animal products (meat, milk, eggs) more important than vegetable ones. For example, GDR dairies in Nauen and Brandenburg provide up to 13 per cent of West Berlin's liquid milk, whereas before the Second World War the former Brandenburg Province provided the bulk of Berlin's milk requirements. Today West Berlin must bring in most of its milk supplies by road tanker from Lower Saxony, Schleswig-Holstein and even more distant parts of the Federal Republic (Reindke 1985). The GDR also provides 60 per cent of West Berlin's requirements of grain for milling and brewing, as well as some higher-value products from the Werder fruit-growing district near Potsdam or from the market gardens of the Spreewald. It is then something of a surprise to find high on the list one commodity, clothing, that does not come under the general rule of heavy or perishable supplies (see Table 2.2).

Among deliveries from West Berlin, iron and steel leads in terms both of weight and of value. Although apparently a curious item for a city not known as an iron and steel producer, it must be remembered that the Tegel special-steel plant refines both scrap and some metal originating in the GDR. More understandable is the prominence, in a slender list, of machinery and of electrical equipment (see Table 2.1).

Mention has already been made of the extraordinary achievement of flying-in equipment during the blockade to rebuild the Kraftwerk

West (later Ernst-Reuter-Kraftwerk), that had been totally dismantled and shipped to the Soviet Union. In March 1952 the GDR entirely cut off West Berlin from its electricity distribution system, but the city has had no difficulty in installing sufficient generating capacity to keep pace with its own needs. The GDR later decided that it was to its advantage to reconnect the two systems, and again sells a small amount of power. By the 1950s West Berlin had also acquired independence in the manufacture and supply of town gas. By 1985, however, under conditions of détente, West Berlin secured a spur-connection from the pipeline supplying Soviet natural gas to the Federal Republic (see section 5.4.2).

Municipal administration in the two parts of the city having been totally divided, contacts remain only on unavoidable technical matters, and these are very few. It is obviously necessary for the West Berlin highway engineers to be in contact with the GDR engineers over alterations to crossing points, such as access to the new Berlin–Hamburg autobahn, or with the GDR-based railway administration when new roads have to be built over or under rail lines. Otherwise what is perhaps the most important area of continuing collaboration relates to water supply and waste disposal. The division of the city caused some difficulties in water supply, particularly for the rather isolated Neukölln District, which depended on a waterworks in East Berlin, but by the 1960s West Berlin could, with minor exceptions, supply itself with water from local resources. West Berlin's liquid effluent, especially from the southern districts, was in the past disposed of by sewage farms and treatment plants in what is now the GDR. It was not in the interest of the GDR to interrupt these arrangements, as any consequent public-health disaster could hardly have been restricted to West Berlin. In the course of time, however, West Berlin has installed new capacity for sewage treatment. With the completion of extensions to its Ruhleben plant, dependence on disposal in the GDR will be reduced to about 25 per cent. Investment by West Berlin on improving the quality of its water bodies, such as the Tegel plant for phosphate reduction, also benefits downstream areas in the GDR. West Berlin has also contributed to the cost of installations in the GDR that improve water quality in the Berlin region as a whole.

Solid waste originating in Berlin was previously disposed of by tipping, mainly in the area of the present GDR. With the isolation of West Berlin, the GDR declined to accept domestic waste, although continuing to accept waste from building operations. Areas for tipping had to be found within West Berlin, but the consequent demands upon the limited open space, and the danger to ground-water supplies, led the city to adopt a policy of incinerating refuse. Under the conditions of détente following the 1971 Berlin Agreement, however, the GDR was willing once more to participate in the disposal of West Berlin waste, at a price in hard currency. Special crossing points for the exclusive use of West Berlin refuse transporters have been created in the east of the city, leading to a disposal point at Vorketzin, and to the south of the city, leading to a disposal point at Schöneiche. With the recent consciousness of environmental problems, West Berlin has been obliged to finance a plant at Schöneiche for the treatment of dangerous industrial wastes. The GDR also continues to deal with building rubble and material from excavations, much of which moves by water.

In addition to the strict administrative division of the city, virtually all institutional links have been broken; artistic, intellectual, cultural, and sporting activities operate in isolated cells. Virtually every institution is duplicated; there are distinct opera houses, symphony orchestras, universities, state libraries, museum systems, even two separate zoos. A West Berlin academic has more chance of talking to his East Berlin opposite number at an international conference in a foreign country than in his own city.

A particular cause of resentment by the GDR has been the question of the custody of works of art, books, and archives that, in 1939, had been in the Prussian state museums and collections in Berlin. In the war years, these collections were in general evacuated to places of presumed safety outside the city. Some items were destroyed by bombing or fighting, and some disappeared in the chaos of the war's end. Surviving items that were retrieved in the territory occupied by the Red Army eventually found their way back to the museums, libraries, or archives of East Berlin. Those that were found in the territory occupied by the western Allies were eventually returned to West Berlin, being placed by federal legislation in the trust of the *Stiftung Preussischer Kulturbesitz*, set up in 1957. Even cultural material belonging to private individuals or societies is affected; to give a minor example, the pre-war catalogue of the famous Berlin Geographical Society is in the state library in East Berlin, which, however, has only a proportion of the books to which the catalogue refers. Some are lost, others are in the new building of the society in Dahlem, West Berlin. The problems for scholars presented by this situation, repeated a hundred times over, can easily be imagined.

The GDR holds that all cultural property originating from its territory, and in particular from the museums and libraries of East Berlin, should be returned to the 1939 location. In 1975 the GDR refused to continue with negotiations for a Cultural Agreement with the Federal Republic until the items had been returned. It also made a practice of refusing to loan items from its collections for exhibitions in other countries which included items from West Berlin. By 1983 there had been a change of heart, and negotiations began on the Cultural

Agreement that is now concluded; the question of the return of cultural objects was 'shelved'. Both sides agreed to do everything possible to minimize the effects of division, for example by exchanges.

At local level in Berlin, something in this direction has already been achieved. Perhaps the most generally acclaimed exchange involved the return from West Berlin of eight marble figures belonging to the Schloss-Brücke (now Marx-Engels-Brücke) in the heart of East Berlin. They have already become a familiar part of the Berlin urban landscape. Among other items, West Berlin has also returned the marble statue of Schiller to stand in its old location in the Gendarmenmarkt (now Platz der Akademie) in front of Schinkel's Schauspielhaus (the state theatre), where many of Schiller's works were given their first performance, and which has been restored as a concert hall. Items returned from East Berlin include the archives of the *Königlich Preussische Porzellanmanufaktur* (KPM), the historic state-owned porcelain works in West Berlin.

The Churches were among the institutions that held out longest against division. The post-war seat of the Berlin/Brandenburg diocese of the Protestant Church was in Charlottenburg, West Berlin. By 1969 a separate 'League of Evangelical Churches in the German Democratic Republic' had been set up, and in 1972 the Protestant Church gave up; a second diocese was established for East Berlin and Brandenburg. The position of the Catholics was rather different; they were neither numerous nor influential, and the seat of the diocese was in East Berlin. Bishop (later Cardinal) Bengsch had been designated but not installed just before the building of the Wall, and was thereafter left free to visit the West Berlin part of his diocese. Once GDR Church organizations had been set up independently of those in the Federal Republic, the GDR government allowed some relations, particularly as the Churches in the Federal Republic were willing to contribute financially to the repair of bomb-damaged churches and to the building of such new churches and church-centres as were permitted.

Merritt used a group of quantitative measures including postal exchanges, visits by East Berlin inhabitants to theatres and similar places of public resort in West Berlin, and letters from East Berlin inhabitants to RIAS, to show that even in the period 1950–61, before the building of the Wall, interaction between the people of West and East Berlin was steadily declining (Merritt 1985). He puts forward a number of explanations for this decline. The movement of 3 million refugees from the GDR to the Federal Republic before 1961 may itself have swept with it many of the people most likely to maintain contacts with West Berlin; even in purely quantitative terms, a lesser intensity of interaction would be expected from a smaller population. It is true that the sector boundary remained open during this time, but West

Berliners may have been deterred from visits by vague fears of trouble with the People's Police, while East Berliners may have feared that a casual control of identity papers on crossing might result in the recording of a 'black mark' that could have adverse consequences, perhaps in relation to employment or access to education. What is perhaps more important is that once the city was divided between two different systems of political organization, there was an inevitable adjustment in patterns of spatial activity. The business of life, such as school attendance, purchasing trips, journey to work (for most people), health provision, and visits to officialdom regarding identity papers, residence permits, housing provision, and the like, all took place within the structure of either West or East Berlin; in such matters a visit 'over there' was an irrelevance. Even in matters of choice, the question of sheer proximity is clearly significant; it is normal to use the nearest shopping centre, to go to the neighbourhood cinema, or to take the children to play in the local park. The sheer passage of time must also not be forgotten. Those who knew undivided Berlin forget, and death year by year reduces their numbers, while birth and migration bring in residents to whom the united city is only a part of history.

The building of the Wall in 1961 of course made interaction between individuals in the two parts of Berlin much more difficult. For the inhabitants of East Berlin the position was, and remains, quite simple; in normal circumstances it is not possible for them to visit West Berlin. Exceptions include persons on official business and persons of retirement age. The latter have no more labour to devote to the economy of the GDR, which would in fact benefit economically were they not to return.

The treatment of West Berlin visitors has varied over time. Initially, they were quite simply unable to visit East Berlin or the GDR, in contrast to residents of the remainder of the Federal Republic or foreigners, who could apply for visas. Very gradually, some concessions were introduced. Beginning in 1963, the GDR issued special entry permits at a limited number of holiday periods. Then from 1964 West Berliners were allowed to apply for visits for urgent family reasons, such as serious illness or death. Merritt has shown how, after a high initial take-up, numbers in both these categories steadily declined (Merritt 1985). West Berliners could, of course, use the transit routes through the GDR, except in periods when the Soviet Union and the GDR decided to obstruct them; the transit visa fees and road taxes imposed by the GDR are now covered by global sums paid by the Federal Republic.

Matters were greatly eased in the period of détente following the 1971 Berlin Agreement. Conditions for West Berliners were somewhat more restrictive than for other citizens of the Federal Republic or for foreigners, but from 1972 they were allowed visits of up to thirty days in the year (even longer in special circumstances such as illness). Increasingly, visiting East Berlin and the GDR has become restricted not by legal requirements but by price, the requirement to purchase a visa and to exchange a minimum daily amount of hard currency. Once again, the number of visits by West Berliners to East Berlin and the GDR has shown a generally declining trend, which may be taken as additional evidence of the way in which interaction between the inhabitants of the two parts of the city is continuing to decline (Merritt 1985). There are very strict limitations on articles that may be taken in and out of East Berlin and the GDR; in particular western books, magazines, and trade catalogues are likely to be confiscated at the crossing points. Berlin travel firms also now offer group tours, not only of East Berlin, but of nearby GDR destinations such as Potsdam.

The postal system, which is governed by international agreements has never been cut off, although items may be opened in search of material in the prohibited categories. Occasionally, also, items of mail may be returned because they bear a special-issue postage stamp deemed unacceptable politically (e.g. stamps issued in 1986 on the occasion of the twenty-fifth anniversary of the building of the Berlin Wall). Postal items from the GDR and East Berlin have to be stamped as if West Berlin were a foreign country, whereas in the reverse direction only internal rates are payable. The number of parcels exchanged with East Berlin and the GDR has dropped off extremely sharply since the mid-1960s. It is impossible to say how much of this decline really reflects a slackening in interpersonal relations and how much it is determined by the improvement of GDR living standards and the fact that it is more economical and safer to transmit money to GDR relatives which, in the form of the special vouchers known as *Valuta* can be used to purchase western luxuries at branches of the special 'Intershop' chain.

The telephone system fared less well than the posts. All direct public telephone lines between the two halves of the city were broken off in the early 1950s, after which there was only a very tenuous operator-controlled connection by way of Frankfurt on Main! Otherwise there remained only a few lines required for the safe operation of the various railway systems or linking the police and fire brigades on the two sides. All this has changed in the period of détente. Direct dialling is now available between the two parts of the city, and the number of lines, although rarely adequate, steadily increased. Merritt has suggested that the use of the telephone may account for some of the decline in the number of visits or postal exchanges (Merritt 1985).

These questions of division and (in respect of West Berlin) isolation will continue to provide major themes in the chapters that follow.

3 | The Berlin countryside

3.1 The evolution of the physical environment

The city of Berlin lies on the Northern Lowland of Germany. This means that the relief features that provide its geographical setting derive almost exclusively from the events of the Pleistocene Ice Age, only slightly modified in the relatively short period of geological time since the ice sheets finally withdrew. Whereas the distant observer tends to picture Berlin as sited in a featureless expanse of pine-clad sandy plain, a closer view reveals a surprising variety of terrain.

Beneath the Quaternary (Pleistocene and Holocene) surface deposits of the last 1.5 million to 2 million years there is a buried landscape of 'solid' geology. As in much of northern Germany, the *Zechstein* (Thuringian) deposits of the upper part of the Permian are of particular interest. In the neighbourhood of Berlin the *Zechstein* is several hundred feet thick and normally lies at a depth of between 1,500 and 2,000 m. The deposits consist mainly of salt, which under pressure has the characteristic of becoming plastic and of forming salt domes, breaching overlying deposits or thrusting them upwards. In the Berlin area a salt dome structure reaches the surface to form the gypsum deposit at Sperenberg, south of the city, while to the east at Rüdersdorf the Triassic Muschelkalk limestone is thrust close enough to the surface to be worked for cement manufacture. Other salt structures are known from relatively shallow borings. Particularly in the neighbourhood of the salt domes, salt springs may reach the surface. In the neighbourhood of Berlin there were from time to time attempts to establish a spa (*Gesundbrunnen*), but to no lasting effect.

The overlying Tertiary deposits are mostly sands, with bands of clay and a thin bed of brown coal. During the blockade, borings were even made to see if brown coal could be worked beneath West Berlin, but the attempt was unavailing. Of great economic importance is the middle-Oligocene bed of clay known as the *Ruppelton* or *Septarienton*, which lies at a depth of between 80 and 100 m below sea-level. This effectively seals off the saline water contained in the earlier Tertiary deposits, thus preserving from contamination the ground water on which Berlin substantially relies for its water supply. In places where this clay has been eroded away, notably on the fringes of salt domes, there is a danger of salt contamination of the ground-water supply (Marcinek, Saratka, and Zaumseil 1983a).

The glacial morphology of the Northern Lowland in what is now the GDR, is fundamentally simple. The pre-glacial surface took the form of a plain descending gently northwards, with the rivers following the same general direction. The continental ice sheet from Scandinavia advanced up this plain, pushing material in front which remains today as terminal moraines, looping across the lowland with a general south-east–north-west direction. Inside the moraines the melting ice left behind plains of boulder clay or till, and outside them melt-water laid down spreads of outwash sands (*Sander*). Because the ice sheets blocked the natural lines of drainage to the north, glacial melt-water together with the drainage from further south had to escape laterally round the outside of the ice, cutting as it did so pronounced pro-glacial stream trenches (*Urstromtäler*), which remain today as a prominent feature of the landscape.

Understandably, the actual position is more complicated than this simplified scheme suggests. In the first place, there was not a single ice advance but a number of distinct glaciations separated by warmer interglacials, with the possibility of additional stages of advance and retreat within any one glaciation. Accordingly, the deposits and morphological features of earlier glaciations or stages may be buried by later ones. Even within one glaciation the retreat of the ice front could produce a sequence of terminal moraines and associated features. This was the case in the closing stages of the Weichsel (Vistula) glaciation, the last north-German glaciation, which left a succession of moraines and *Urstromtäler*. The latter are in part still occupied by the rivers, which however periodically break through the moraine wall to the next *Urstromtal* to the north, producing a characteristic rectangular pattern of drainage.

The countryside of Berlin is essentially the product of the Weichsel glaciation. Underlying deposits derived from the earlier Elster and Saale glaciations are nevertheless important, because the buried outwash sands that they contain are valuable aquifers. About 20,000 years ago the Weichsel ice sheet was at its greatest extent, forming the Brandenburg moraine system 40 to 50 km south of the city, with on its outer side the Baruth *Urstromtal*. The advance of the ice to this point contributed to the excavation of basins which, filled with dead ice that gradually melted in warmer climatic conditions, formed the Berlin lakes. The advancing ice only partially remodelled the previous relief; in some parts features from previous glaciations remained with only minor modification.

The ice sheet then retreated by stages, over a period of perhaps 2,000 to 3,000 years, to form another major moraine system, the Frankfurt moraine, north of the city. During this process there was further excavation of the basins that would emerge as part of the lake system around the city. The retreating ice left behind its quota of till on the Teltow plateau south of the city, which assumed very much its present contours. At the time of the formation of the Frankfurt moraine, the melt-water streams emerging from the ice were gathered up in the Berlin *Urstromtal*. With the further northern retreat of the ice front, the Berlin *Urstromtal* continued to receive melt-water, the channels emerging from the ice defining the Barnim plateau of glacially derived material north of Berlin (Marcinek, Saratka, and Zaumseil 1983a).

3.2 Landscape types of the Berlin region

3.2.1 The Urstromtal

The core of the city of Berlin lies in the Warsaw–Berlin *Urstromtal*, which also contained two other important medieval towns, Spandau and Köpenick, now both engulfed in Greater Berlin (fig. 3.1). The floor of the *Urstromtal* lies between 30 and 40 m above sea-level; it is at its narrowest at the site of Berlin. Its floor is composed of a 40–55 m thickness of sands and gravels of both Weichsel and Saale age. The ground-water contained in these sands is the principal basis of Berlin's water supply; it is derived not only from the direct contribution of rainfall but from the inflow of water from glacial sands incorporated in the higher ground to north and south, notably Barnim.

The *Urstromtal* in the Berlin area is occupied by the Spree. Upstream it is joined by the Dahme, downstream it enters the Havel. The fall of the *Urstromtal* floor, both longitudinally and transversally, is extremely slight; to the east the surface of the Müggelsee lies at 32 m above sea-level, while the Havel between Spandau and Potsdam lies at 30 m, a minimal fall (Leyden 1933). The Spree responded by meandering and by splitting into braided channels, as at the initial site of Berlin. Particularly since the middle of the nineteenth century, the stream has been straightened, and regulated by means of weirs, and equipped with locks for river traffic. The abandoned branches of the river have in part been utilized to form branch canals and harbours for industrial or public-utility plants, in part filled in for building purposes. The lack of fall led to extremely swampy conditions along the flood plain of the Spree, with considerable peat accumulation. Under-

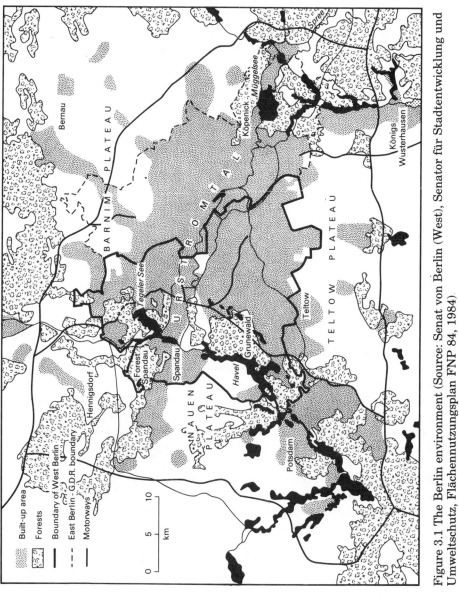

Figure 3.1 The Berlin environment (Source: Senat von Berlin (West), Senator für Stadtentwicklung und Umweltschutz, Flächennutzungsplan FNP 84, 1984).

standably, this has created building difficulties for central Berlin, which lies across the flood plain. Except where, like the ancient Berlin churches, they can take advantage of former sandbanks, buildings have to be supported on piles, or other special measures taken to secure their foundations. Away from the densely built-up core of Berlin, much of the Spree flood plain is used for river ports and other transport installations, industrial or public-utility sites (Niederschöneweide, Siemensstadt, Spandau), parks (Schloss Charlottenburg, Treptow), or allotments (Spandau).

3.2.2 The glacial-till plateaus

Even within the closely built-up area of Berlin it is possible to observe a marked increase of gradient in streets running north or south from the inner city, such as the Schönhauser Allee or the Prenzlauer Allee, mounting northwards into the significantly named Prenzlauer Berg District. These breaks of slope mark the transition to the second major relief type of the Berlin region, the low glacial-till plateaus of Barnim to the north and of Teltow to the south. In absolute terms the difference in elevation is not great; the Kreuzberg, the well-known viewpoint on the northern edge of Teltow in Kreuzberg District, rises only to 66 m, about 30 m above the level of the *Urstromtal*. From the late nineteenth century onwards, Berlin has expanded on to the plateaus, with the present West Berlin taking in a large part of the Teltow Plateau, and with East Berlin still advancing over Barnim. Further glacial-till plateaus are to be found outside the present city limits, notably the Nauen plateau to the west.

Essentially, the plateaus are characterized by rather featureless spreads of unsorted boulder clay or till, left by the decay of the last of the ice sheets. The till covers some scores of metres of sands and bands of clay deposited by the advancing Weichsel ice and by earlier glaciations, forming important aquifers. In places the retreating Weichsel ice also spread relatively thin sands over the till. In the course of human occupance the clay areas have been cleared for arable, the sands often left under forest. Soil conditions even explain some aspects of the development of the modern city; the sector Lichtenberg-Marzahn-Hellersdorf, although relatively close to the city on its eastern side, long repulsed communications and settlement because of its heavy soils, being finally opened up only by the major East Berlin housing projects of recent years. The Teltow plateau presents the most uniform till surface, although varied by some terminal-moraine and fluvial-glacial features east of the Havel (the Grunewald Forest).

Barnim is rather more complex, as it was crossed by melt-water streams from the Frankfurt moraine, heading for the Berlin *Urstromtal*. These have left sand spreads covering the till along the plateau's western fringe and, more particularly, further east in the neighbourhood of Strausberg.

The till plateaus are characteristically diversified by a number of minor glacial-relief features. Fissures in the glacier, occupied by melt-water laden with sand and gravel, survived the melting of the ice in the form of eskers, elongated features snaking across the countryside, while kames originated as steep-sided alluvial cones built up by melt-water emerging high on a glacier. Such ice-margin features combine with terminal-moraine remnants to form the hills that rise above the general Teltow surface on the edge of the Grunewald, overlooking the Havel, reaching 97 m in the Karlsberg.

The movement of melt-water could also scoop out narrow, steepsided sub-glacial channels (*Rinnen*), which are not now occupied by continuous drainage. Characteristically, they contain small elongated lakes or boggy areas where lakes have been filled. One of the most remarkable is the Grunewald channel, which can be traced as a series of lakes and bogs from the park of Schloss Charlottenburg in the north southwards through the Grunewald forest to the Havel north of Potsdam (see fig. 3.1). The course of another such valley, on the southern fringe of the city, is now substantially followed by the Teltow Canal. Within the built-up area some of the lakes have been retained as features in parks, but others were filled in by tipping, often leaving subsequent buildings with foundation problems. The retreat of the glacier left masses of ice buried in the till, which melted only slowly as permafrost gave way to subsequent warmer conditions. The resultant kettle-holes, when filled with water, formed ponds studding the Berlin area. Most have by now been filled, either by natural sedimentation or human interference, but some remain. One particularly well-known example is provided by the pond round which the architect Bruno Taut built the *Hufeisen* (Horseshoe) flats, perhaps the masterpiece of Berlin social housing of the 1920s.

Characteristic of the till plateaus is the profusion of roughly rounded stone fragments derived from Scandinavia, grading upwards in size from fluvio-glacial gravel to true erratic blocks of a metre diameter or more. The larger blocks had to be removed from the fields to allow cultivation and, in an area otherwise lacking in stone, are to be found forming the walls of the oldest churches. Such stones were also used in rural areas for surfacing roads, providing a punishing experience for drivers until replaced by modern materials. Particularly large erratic blocks were used for war memorials and other monuments.

3.2.3 The system of major lakes and rivers

A characteristic of the built-up area of Berlin is the presence on its western and eastern fringes of considerable systems of lakes. These are generally held to have originated as channels used by melt-water either under or in front of the ice sheets, possibly on the lines of the pre-glacial south–north river drainage system. They are clearly more substantial and older features than the sub-glacial channels that score the boulder-clay plateaus. The north–south trending Havel channel, for example, appears to have been excavated to a depth of 23 m below sea-level in the Elster glaciation, and was further scoured by glacial and fluvio-glacial action during the succeeding Saale glaciation, and probably during the Weichsel glaciation (Pachur, Schulz, and Stäblein 1985). A generally north–south orientation, parallel to the direction of ice advance, and to the sub-glacial evacuation of melt-water, is also characteristic of the lakes to the east of Berlin. It is likely that these channels were preserved from silting well into the post-glacial warm period by being filled with dead ice, that melted only slowly. Ice-margin moraine and kame features contribute to the close association of the lake systems with the highest (if still modest) elevations in the whole of the Berlin region, reaching, as noted, to just under 97 m in the Grunewald in West Berlin and to 115 m in the Müggelberge in East Berlin. This degree of relief energy contributes to the attraction of the lake areas for tourist purposes.

The drainage network in the Berlin area is relatively immature, closely consequent on the erosional and depositional features of the Weichsel glaciation. The overall slope of land in the Northern Lowland of Germany being northwards, towards the Baltic, it is not surprising that the Spree upstream of Berlin (once it breaks out from the Baruth *Urstromtal*) and its major right-bank tributary, the Dahme, flow in this direction, but they are then gathered up by the Berlin *Urstromtal* to turn towards the west. The drainage of Teltow and southern Barnim is generally southwards, so that the Spree is essentially restricted to right-bank tributaries in its course through the city, including the Panke, now to be glimpsed only as a muddy ditch in its generally culverted southern course. At Spandau the Spree joins the Havel, which, contrary to the general rule for rivers of the Northern Lowland, is at this point flowing to the south. It is possible that the river was pulled into this southerly course after the close of glacial times, when dead ice melted to reveal the deeply scoured Havel lakes. A further oddity of the rivers of the Berlin area is the way in which the Havel–Spree system drains a strip of territory embracing nearly the whole of the eastern side of the GDR, from the Czechoslovak boundary almost to the Baltic, but then escapes by a convoluted route to the Elbe and the distant North Sea, instead of joining the Oder, only a few score kilometres away to the east. The Elbe–Oder watershed actually runs across the Barnim plateau northeast of Berlin.

The oddities of the river pattern, with extremely low watersheds between present-day streams provided by the *Urstromtäler* and *Rinnen*, together with the numerous lakes, have greatly facilitated canal construction and waterway improvement.

3.2.4 Post-glacial (Holocene) features

The formation of the landscape did not end with the withdrawal of the ice. The land surface lay at first unprotected by vegetation, so that winds picked up the silty particles in the drift to deposit them far to the south as loess, dropping the coarser paticles to form dunes, particularly in the *Urstromtal*, and to a lesser extent on the plateaus. Dune sands are too shifting to be used for building, so that dunes within the built-up area are characteristically used for cemeteries and similar purposes. Outside the built-up area they are usually forested. The withdrawal of the ice sheets was immediately followed by a period of periglacial evolution. In this time, slopes were modified, while the existence of permafrost led to the development on the plateaus of polygonal soil structures and frost wedges.

With the arrival of the warmer climatic conditions of our own time, additional deposits were mainly limited to muds and peat accumulated on the river flood-plains, in lake beds and in other depressions with a high water table. Such deposits are found filling in many of the kettle-holes and sub-glacial channels of the plateaus, but the most extensive accumulations were naturally along the rivers and in the vicinity of the lakes. Generally, deposits are thin, but locally, where dead ice masses had decayed or there had been scouring by ice or melt-water streams well below the present general land surface, thicknesses can be considerable, for example 48.6 m at the East Berlin 'Museum-Island' (Marcinek, Saratka, and Zaumseil 1983a) or 30 m of lake deposits at the Havel (Pachur, Schulz, and Stäblein 1985). As noted above, the presence of such peaty, unstable material along the *Urstromtal* through Berlin was a source of considerable difficulty for building operations in the development of the city.

3.2.5 Human intervention in the landscape

The expansion of the urban area of Berlin led to the extinction of many ponds and minor lakes, as well as to extensive tipping of material to fill up sub-glacial channels or to raise the Spree flood plain in order to provide building sites. Buildings and other structures built across infilled channels or ponds have in places been affected by differential settlement, even leading (as at the northern end of the Grunewald channel) to the eventual demolition of buildings. Where ponds or lakes have been retained, for example as features in parks, they have often been dredged and 'landscaped'. As already noted, the course of the Spree has been straightened and regulated, with the removal and generally the infilling of former meanders. Some lesser streams, such as the Panke, have disappeared altogether over sections of their courses, while canal-building and harbour construction has further modified the pattern of waterways (see section 4.1.1).

A characteristic of Berlin, springing from one of the darkest periods of its history, is the presence of *Trümmerberge* (rubble hills), where the rubble from the bombed city was tipped once it had been sifted for reusable bricks and other material. In West Berlin such tips were also used in the years of maximum post-war isolation for the disposal of domestic rubbish. Sometimes a tip was built round a *Hochbunker*, one of the massive concrete structures that provided both air-raid shelters and an emplacement for anti-aircraft artillery. Generally, an attempt was made to site the tips where there was already a varied terrain into which they could be landscaped, as in Volkspark Friedrichshain on the hilly edge of Barnim, where the *Bunkerberg*, or as it is unofficially known 'Mont Klamott' (Rubbish Mountain) provides an excellent view over the Berlin inner city. Perhaps the most dramatic use of rubble was to raise the Teufelsberg in the Grunewald, already among the highest elevations in West Berlin, to a height of 114.7 m. Today it is crowned by a military radar installation that is one of the most widely visible features of West Berlin. A more recent example of large-scale creative landscaping is provided by the hills and lake formed on the flat till surface of the Teltow plateau in southern Neukölln for the 1985 Federal Garden Exhibition; the features will remain as part of a public park.

3.3 Water

3.3.1 Water supply

At the beginning of the nineteenth century the 300,000 inhabitants of the city relied for their water upon wells or pumps set up in the streets and courts. The old street pumps can still be seen in the older parts of the city, sometimes reused as decorative features, but still capable of providing water for such purposes as car-washing. In the *Urstromtal*, to which the urban area was still essentially confined, water could be reached at depths of only 2 to 3 m, but unfortunately cess-pits for sewage disposal were usually sunk in close proximity. Waste water was also thrown into the street gutters, making its way slowly to the rivers, so there were plentiful opportunities for disease transmission. The first water-supply system was developed by a British firm in 1852–6; it drew water from the Spree at Stralau, just upstream of the then built-up area, and pumped it to a waterworks on the fringe of the Barnim plateau, from where it was distributed by gravity to a proportion of the houses of the city; the remainder continued to rely on pumps and wells.

In 1873 the city took over the task of water supply, and over the next thirty years developed the basic structure of the supply system as it now exists. Instead of the drawing of water from the Spree at Stralau, water-supply operations were transferred to more distant sources in the *Urstromtal* above and below the city, the two major sources being the Müggelsee upstream and the Tegeler See which, although west of the city, lay above the point at which the polluted waters of the Spree entered the Havel. There were various stages in the evolution of the system, including the supply of untreated ground-water, then the supply of treated lake water (the first treatment plant was the present Friedrichshagen plant on the Müggelsee, opened in 1893), until the present basic reliance on treated ground-water was introduced in 1900 (Marcinek, Saratka, and Zaumseil 1983a: 28–33). The principle of concentrating the greater part of water-abstraction and treatment plants in the *Urstromtal* east and west of the city has held to this day, although naturally the number of plants has increased, and existing plants have been extended from time to time.

The basic method of ground-water abstraction is by bank-filtration; this involves the sinking of wells close to lakes and rivers. The water this yields derives in part from the rivers that flow into Berlin, in part from the part of precipitation that reaches the aquifers beneath the city. In the built-up area of Berlin the amount of the year's precipitation reaching the aquifers is extremely limited; the majority runs off from roofs and roads into surface-water drains and eventually into the rivers. Even areas which are not built-up are often floored with boulder-clay, producing run-off rather than infiltration. It has been estimated for West Berlin that only between 10 per cent and 15 per cent of precipitation enters the water table (Treter 1985). Fortunately, ground-water supplies are reinforced by the underground movement of water that has infiltrated outside the city, notably in Barnim.

The derivation of Berlin's water supplies from distinct upstream and downstream areas of supply meant that, on the division of the city, although there were some problems with the distribution networks (see section 2.7) West and East Berlin each received a share of resources. As in all advanced societies, improvements in household equipment have led to a marked increase in demand for water; in West Berlin, for example, domestic demand increased from 91.7 million cu. m in 1963 to 115.4 cu. m in 1981. Supplies by the public water-supply organization for industrial and commercial purposes also increased, but by nothing like the same proportion, from 56.3 million cu. m to 66.4 million cu. m over the same period. It is not clear to what extent this reduced rate of increase reflected improved industrial practices, for example in the use of cooling water, or to what extent it reflected changes in the structure of Berlin industry. The water-supply company was able to meet the increased demand, but obviously only by drawing more heavily on the available ground-water, a demand only partially offset by a steady reduction in water directly abstracted by industry and public utilities. In addition, ground-water is lost through pumping operations in the course of building and construction projects. There is no doubt that, at least until very recently, ground-water has been removed at a faster rate than it can naturally be replenished, leading to a general fall in the water table; for example, a drop of up to 5 m is recorded for the Grunewald area between 1921 and 1978 (Treter 1985).

Efforts are now being made to replenish ground-water with surface water. To a certain degree this happens automatically; the lower the water table sinks, the more water is likely to seep from the waterways into the wells sunk for bank-filtration. This source is now supplemented by the deliberate reintroduction to the aquifer of water derived from such sources as building operations or urban run-off. Gains from artificial replenishment, coupled with an economically determined decline from the 1970s in losses through building operations and industrial abstraction, meant that by the early 1980s the decline in underground reserves had been checked, and there were even signs of a slight rise in the water table (Treter 1985).

Developments in East Berlin have been parallel; there has been the same soaring demand for water in a city where industry and population are expanding, and a major programme of rehousing to modern standards is in progress. The main supplies are drawn from ground-water in the upstream *Urstromtal*, but at times of high demand water is also taken directly from the Müggelsee. A number of the waterworks have had to be expanded to meet increased demand, and as in West Berlin, attention is now being focused on the artificial enrichment of ground-water. As East Berlin is not separated from its surrounding territory, it is possible for water to be drawn from further afield. One proposal is to draw water from the Oder by way of the Oder–Spree Canal, benefiting from the self-cleaning capacity of waterways in the course of transfer (Marcinek, Saratka, and Zaumseil 1983a).

3.3.2 Effluent disposal and water purity

At much the same time that Berlin was installing its modern water-supply system, it began to install a proper drainage system. The customary method of liquid effluent disposal was by spreading in the shallow lagoons (*Rieselfelder*) of a sewage farm (*Rieselgut*), situated somewhere outside the built-up area. It has already been noted (see section 1.3) that the division of the Berlin metropolitan area before 1920 between a number of separate local authorities led to a multiplicity of such operations. As the sewage farms lay mainly outside the territory of Greater Berlin, the 'island' of West Berlin after the division of the city found itself heavily dependent on the GDR for effluent disposal (see section 2.7). The provision of modern effluent-treatment facilities, as at Ruhleben on the Spree above Spandau, reduced dependence on the GDR to about 25 per cent, but this was not the only reason for change. In both West and East Berlin, effluent disposal by means of sewage farms fell out of favour in the post-war period. While the method did something to enrich ground-water supplies, it also carried the dangers of pollution, the accumulation in the soil of heavy-metal residues and possibly the spread of disease. The use of modern treatment plants also means that the considerable areas occupied by the sewage farms can be turned over to other forms of land use.

Following an agreement signed between the government of the Federal Republic and the GDR in 1982, both sides have co-operated in a common attempt to improve water quality in the Berlin area. West Berlin attention has focused particularly on the condition of the Tegeler See, which is not only the source of a third of Berlin's drinking water but a highly valued recreational resource. The lake had become overloaded with phosphate, derived from detergents and agricultural fertilizers used in the area draining into it. The consequence has been excessive growth of algae and decline of oxygen, with highly adverse consequences not only for water quality but for the entire biological environment and recreational amenity of the lake.

A first attack on the problem came in 1980 by floating turbine blowers on the lake to raise the deeper waters to the surface and charge them with air. A more dramatic development was the opening in 1985 of what was at the time the largest plant in the world for the elimination of phosphate from water, intended over the course of five to seven years to return the lake water to the quality it possessed at the beginning of the century. Associated measures include the sewering of settlements around the lake and provision for treating the resultant effluent. As well as improving the quality of the Tegeler See, the plant will also benefit the chain of Havel lakes downstream, the shores of which are shared by West Berlin and the GDR.

The federal government has also contributed to the cost of building the new effluent treatment plant Klärwerk Nord at Schönerlinde, Kreis Oranienburg, in the GDR just north of the city boundary, which with two other new plants under construction in the GDR will contribute greatly to the improvement of water quality in the Berlin area generally. The GDR continues, of course, with other measures of its own, reducing the number of polluted discharges into watercourses, extending the area of the city served by sewers (nearly a third of the built-up area; the less densely built-up part, was still not sewered at the beginning of the 1980s) and linking the sewers to the new or extended treatment plants (Marcinek, Saratka, and Zaumseil 1983b). Neither East nor West Berlin can do much about the fact that the Spree arrives in the city already heavily polluted from the Niederlausitz industrial area upstream.

3.4 Berlin's climate

As seen from Europe's oceanic fringe, Berlin is perceived as having a starkly continental climate, with cold, sunny days in winter when the Siberian blast sweeps over a snow-clad surface, and hot summers tempered by the occasional thunder-storm. More scientifically, Berlin's climate is a transitional one, in which East European continental and West European maritime influences battle for supremacy, although obviously the periods of 'westerly' influence are shorter, the periods of 'continental' predominance longer, than on the oceanic fringe. Berlin

Table 3.1 *Selected climate statistics for Berlin*

	Jan.	Feb.	Mar.	Apr.	May	June	July	Aug.	Sept.	Oct.	Nov.	Dec.	Year
(1) Mean daily length of sunshine (hours)	1.7	2.5	3.6	5.3	7.3	7.5	7.0	6.4	5.1	3.4	1.9	1.3	
(2) Mean daily length of sunshine as percentage of astronomically possible duration	20	25	31	38	47	45	43	44	40	32	33	17	
(3) Number of days with cloud	17.1	14.4	12.4	9.8	9.0	9.1	10.1	8.6	8.3	11.8	15.6	17.4	143.6
(4) Mean monthly and annual temperatures in °C, 1909–60	−0.3	0.1	3.7	8.5	13.7	16.6	18.5	17.5	13.9	8.9	4.1	1.1	8.9
(5) Mean monthly and annual rainfall, mm	48.9	36.8	33.0	42.8	48.9	62.4	75.9	66.4	45.8	46.6	46.9	47.6	601.0

Source: Marcinek, Saratka, and Zaumseil 1985.

is, in fact, sufficiently 'maritime' to have temperatures which, especially in winter, are well above the average for its latitude.

Temperature depends in the first instance on the quantity of radiation received from the sun. This will be reduced in winter by the low angle of the sun and the shortness of the day with, by contrast, the highest figure in June (see Table 3.1(1)). A very low proportion of the astronomically possible sunshine is actually experienced at the earth's surface, especially in winter (2), the variation reflecting the degree of clouding (3). This is at its maximum in winter with a secondary maximum in July. Clouding is at its minimum in September, with a secondary minimum in early summer. The degree of clouding is very much related to the major weather types experienced. Westerly weather brings a rapidly changing situation, cloud and rain being associated with the passage of fronts. Continental anticyclonic conditions bring stable weather and skies that are often clear for long periods. The comparatively cloud-free months of September and June can be related to the predominance of anticyclonic conditions.

The air masses associated with these main weather types also have a direct impact through the temperatures that they import. Average monthly temperatures reach their minimum in January (− 0.3°C), but the absolute January minimum can vary from year to year between − 2°C (when maritime conditions have tended to prevail) and as low as − 25°C (under extreme continental influence), although in most years the absolute minimum varies in the range − 5°C to − 15°C (Leyden 1933: 14). The mean monthly range of temperature is even greater in February than in January, reflecting a more intense battle between maritime and continental conditions.

Summer temperatures have their maximum in July (mean 18.5°C), but compared with winter, variations relating to the battle of maritime and continental climates are much less; the sea, moderately warm, has much less of an impact in reducing continental summer temperatures than in increasing winter ones. This is reflected in the low mean monthly range figures in summer. Absolute July maxima rarely stray very much from 30°C. Leyden, quoting a hundred-year series 1830–1930, found that the maximum in July and August had never been below 25°C, and only on five occasions over 35°C (Leyden 1933). The limited variation in the July figures also reflects the continuity of anticyclonic dominance at this time of the year.

The progress of temperature change through the year shows a marked tendency to 'lag behind the sun', with figures for July to October markedly higher than for the corresponding months in spring and early summer (3.2). This reflects the way in which land and water bodies, once warmed, release that warmth only slowly. The frequent presence in autumn of anticyclonic conditions with warm air masses is also significant (Marcinek, Saratka, and Zaumseil 1983a: 40).

The importance of the oceanic contribution to Berlin's climate is underlined by the figures for predominant wind direction; 34 per cent of the wind blows from the south-west, west and north-west. East winds, which blow much less frequently, tend however to be particularly prevalent in months such as February and March, when they can be very cold indeed.

The rainfall of Berlin totals slightly under 602 mm in the year, with distribution throughout the year such that shortage of rainfall is rarely a problem. The contesting continental and oceanic influences upon Berlin's climate are reflected in the annual distribution of rainfall. The primary monthly mean rainfall peak (76 mm) for July is predominantly of continental type, reflecting the high summer humidity on the continent, the July peak of winds of north-west origin, and the strongly developed summer convectional currents and thunder showers (Marcinek, Saratka, and Zaumseil 1985a: 43). There is, however, a secondary 'oceanic' peak in January, reflecting the winter maximum in the frequency of precipitation. March–April and (to a markedly lesser extent) September, with more settled and frequently anticyclonic conditions, provide two periods of minimum precipitation (Marcinek, Saratka, and Zaumseil 1983a: 43). Within Berlin, rainfall totals reach 620 mm in the higher western areas of the Grunewald and Tegel forests, drop to below 540 mm along the *Urstromtal*, and then rise again on the southwest edge of the Barnim plateau.

Just as the summer maximum of precipitation reflects the continental rather than the oceanic aspect of Berlin's climate, so too does the relatively long average number of forty-three winter days with snow cover. However, the periodic arrival of oceanic air is reflected in the fact that the period of snow cover is rarely continuous (Marcinek, Saratka, and Zaumseil 1983a: 43). The period of snow cover and the quantity received varies greatly from year to year; Leyden records that in the exceptionally severe winter of 1928–9, uninterrupted snow cover lasted for seventy-two days, from 31 December to 12 March (Leyden 1933). In this winter, the Berlin street cleansing department had to remove 1,061,105 cu. m of snow, as compared with only 3,593 cu. m in the following year!

Like all great cities, Berlin to some extent produces its own urban climate. The well-known 'heat island' effect operates all the year, but it is particularly apparent in still weather in winter. Causes of the anomaly include the effects of human activities (heat emissions from vehicles, domestic heating and industry) and also the 'greenhouse' effect, whereby polluted, dust-laden air reduces radiation when a mass of cold air covers the city. Taking mean annual figures, the difference between central Berlin and Potsdam is usually just over 1°C, but in extremely cold conditions the difference can be more spectacular. Leyden quotes an extremely cold day in January 1893 when the lowest recorded street temperature in Berlin was − 23°C, while on the edge of the city in Spandau it was − 31.5°C and in rural Blankenburg nearly − 32°C (Leyden 1933). The heat-island effect also reduces the number of days with ice and frost; plants in the city develop more rapidly in spring than those in rural areas. Particularly in summer, the additional urban warmth increases convection, resulting in a rainfall in the Berlin area perhaps 5 per cent greater than would otherwise be expected (Marcinek, Saratka, and Zaumseil 1983a: 43–7).

3.5 Atmospheric pollution

Berlin, like any great metropolitan area, is subject to severe atmospheric pollution. The principal source of pollution is the large expenditure of energy by industry, domestic heating, and transportation within the city itself. Whether the sources of such emissions are located in East or West Berlin does not make a great deal of difference; pollution is something that the two sides of the city share. The city's own 'heat island' is also a factor. When cold air covers the city, heating from below produces a temperature inversion, which traps air pollutants close to the ground, resulting in the build-up of smog. In the first few days of February 1987, for example, smog conditions were so severe that private cars were banned from the streets of West Berlin, and industries required to reduce production. The average year has seventy-four days with a temperature inversion situated at 300 m or less above the city, forty-one of them in winter, followed by nineteen days in autumn and eleven days in spring (Ergenzinger 1985). Berlin also has the disadvantage that, particularly under winter anticyclonic conditions, when wind direction is from south or south-east, it imports pollution from the GDR industrial areas of Saxony and Niederlausitz, and even from as far afield as Bohemia and southern Poland. These sources can, at times, provide 50 per cent of total pollution.

The most obvious form of urban air pollution is sulphur dioxide (SO_2), which is derived from the burning of coal and oil. It is likely that in Berlin before 1913 SO_2 pollution was much greater than today, because of the reliance not only of households but of numerous small industries upon individual stoves and furnaces. Certainly, it has been only in very recent years that estimated SO_2 emissions per head or per unit area in West Berlin have approached the estimated levels for Greater Berlin during the economic boom of the late 1920s, in spite of a much greater consumption of energy.

There has, however, been a marked shift in the source of such pollution. When West Berlin emerged from the Second World War

(1951), about half of SO_2 emission was still to be attributed to domestic heating and to other small users, and only a quarter to public power stations. By the end of the 1970s total emission had approximately doubled, but the share of domestic and other small sources had diminished absolutely and relatively (19 per cent), although their energy consumption had increased. The share of power stations by contrast had risen to 70 per cent (Karrasch 1985). In spite of the increase in total emissions, the change in origins was a significant gain for the Berlin environment, as domestic sources characteristically vent at a low level, with direct environmental impact, whereas the emissions from the tall chimneys of power stations are, in normal conditions, carried outside the city. Considered spatially, SO_2 pollution is at its greatest in the Wilhelmian ring, in working-class Wedding to the north of the inner city, and Schöneberg and Kreuzberg to the south. It can be related to the age of the housing stock in these districts, with its heavy reliance on individual domestic stoves for heating, all too often fuelled with high-sulphur brown-coal briquettes (Karrasch 1985).

A quantitatively lesser but significant degree of pollution by gases derives from the ever-increasing numbers of motor vehicles. Emissions include carbon monoxide, hydrocarbons, and nitrous oxides, as well as dangerous elements such as lead and cadmium. Pollution from vehicles is at its maximum in the inner parts of West Berlin and along the inner-urban motorways. It is unlikely that federal regulations requiring a reduction in the lead content of petrol and the sulphur content of diesel fuel will do more than hold this type of pollution to its present level, in the face of a continuing increase in the number of motor vehicles.

In addition to various gases, atmospheric pollution in West Berlin includes a quantity of particulates (dust) that is greater than for other major cities of the Federal Republic. Once again, concentration is high in inner districts such as Kreuzberg and Schöneberg, as well as along major road arteries and in proximity to industrial areas both within West Berlin and immediately across its boundary.

Air pollution in Berlin is reflected in damage to the stonework of buildings and monuments, and also in 'acid rain' impact on trees. Research in West Berlin suggests that mortality data for lung cancer, cardiovascular diseases, and respiratory diseases can be correlated with the incidence of air pollution (Karrasch 1985).

It is unlikely that the situation in East Berlin is very different from that of West Berlin. SO_2 concentrations ten to forty times higher than normal and dust concentrations ten to fifteen times higher are reported as characteristic. As in West Berlin, the greatest air pollution is found in the central districts, with the fall-off towards the exterior broken at points of industrial concentration such as Lichtenberg. On the fringe of the city, a local point of high pollution is provided by the cement industry of Rüdersdorf. A particularly characteristic contribution to East Berlin pollution is provided by the two-stroke engines fitted to most East Berlin cars (Zaumseil, Marcinek, and Saratka 1983).

3.6 Vegetation and land use

A common mental picture is that Berlin is a city set in a ring of sombre pine forests, growing out of a sea of pure sand. Reality is a little more complex. 'Scots' pine (*Pinus sylvestris*, the German *Kiefer*) was indeed an early colonist after the final withdrawal of the ice, but with the coming of warmer conditions a sequence of deciduous trees followed: hazel (*Corylus*), lime (*Tilia*), elm (*Ulmus*), ash (*Fraxinus*), maple (*Acer*), oak (*Quercus*), finally beech (*Fagus*), and hornbeam (*Carpinus*). The earliest human occupants of the Berlin area discovered an area of predominantly deciduous forest. Stands of pine were restricted to particularly dry areas, such as dunes or spreads of fluvio-glacial sand, as along the Dahme river south-east of the city. The continuity of the forest cover was only broken by marshes fringing the water bodies and by areas of peat bog (Marcinek, Saratka, and Zaumseil 1983a).

It is now virtually impossible to find any surviving vegetation that can be called 'natural'. Outside the built-up area, Berlin today consists of a mosaic of cleared land and of forest; both are the result of human activities. There have been a number of distinct stages of clearance. Slav clearances along the *Urstromtal* and other valleys were followed by a massive expansion by the German colonists in the twelfth and thirteenth centuries, particularly concentrated on the till plateaus. A further onslaught on the natural vegetation came in the course of the planned creation of new settlements in the second half of the eighteenth century, while the food demands of an increasing population and the discovery of artificial fertilizer led to some additional areas being taken into cultivation in the nineteenth century. The clearance process was an experimental one; between each advance there were periods of adjustment and retreat, when land that had proved unrewarding was given up. It has been suggested that only about half of the wooded area around Berlin has been continuously tree-covered, the remainder having been agriculturally used at some time or other. From the eighteenth century the introduction of scientific forestry led to the increasing development of pure stands of pine. The tree was considered ideally suited to the poor, dry, sandy soils that were not wanted for agriculture but, as is now realized, contributed to their further impoverishment through leaching, podsolization, and the development of acid condition. In recent years there has been some increase in the planting of broad-leaved trees, particularly in recreationally important areas.

Forest is the predominant form of land use in the environment of Berlin, covering 45.6 per cent of the land area of the five immediately adjoining GDR *Landkreise*. The boundaries of the Greater Berlin of 1920 were drawn sufficiently widely to include considerable areas of forest, making up 16–17 per cent of the area of both East and West Berlin, where it constitutes a precious recreational resource for the island city. Areas of loam and brown-earth soil on the till plateaus are generally under arable cultivation; agricultural land makes up 39 per cent of the area of the five immediately adjoining *Landkreise*. Agricultural and garden land even makes up about a quarter of the area of East Berlin (1979), particularly in its north-east sector on the Barnim plateau, although it is shrinking under the impact of new housing projects. There is even a carefully protected 7 per cent of agricultural and garden land in West Berlin.

3.7 Environment and recreation: West Berlin

3.7.1 *Short-period recreation*

The inhabitants of any big city normally make numerous short-period recreational trips, for a day or half a day at a time, up to 100 km deep into the surrounding countryside, especially when it is as attractive as the forests and lakes surrounding Berlin. In 1952 this possibility was effectively removed from the inhabitants of West Berlin. It is true that after the 1971 Berlin Agreements the possibility of visiting the GDR and East Berlin was restored to West Berliners, but visits are effectively discouraged by the need to apply in advance for permission to make up to thirty day-visits in the year, by the requirement to exchange a minimum sum of foreign currency for each day, and by the intensive and time-consuming controls at each crossing of the boundary. Such visits as are made are usually to visit relatives, not for recreational reasons (Steinecke 1985).

One effect of West Berlin's 'island' situation is for recreational trips in the local environment to be replaced by short-period tourist visits to the Federal Republic, but the expenditure of time and money involved in such trips is clearly considerable. By autobahn the nearest crossing point, at Marienborn/Helmstedt is 186 km, with the crossing points in the direction of Hamburg, northern Bavaria and Hessen progressively further away, to which must be added whatever distance is travelled inside the Federal Republic. What is more, the West Berlin driver must endure the frustration of having to throttle back on the autobahn to the sedate GDR maximum speed of 100 km (62 miles) per hour. The passing of two border control points will add anything up to another half hour to the journey, even several hours at holiday periods. Understandably, with this additional hurdle, fewer Berliners take this type of short holiday than inhabitants of the large cities of the Federal Republic, such as Hamburg. Those who can make the long journey, who can afford second homes or hotel rooms in the Harz or the Lüneburger Heide, are predominantly the higher-income people. Those who are already disadvantaged, the old, the foreign workers, the German unskilled workers, and the very large retired population, also have a diminished possibility of taking holiday spells outside West Berlin. This is why it is official policy to maximize opportunities for recreation within the city's boundaries; moreover every addition to recreational possibilities makes the city more attractive to the young and skilled people that the authorities hope to attract from the remainder of the Federal Republic (Steinecke 1985).

3.7.2 The West Berlin forests and lakes

Life in the island of West Berlin would be a great deal more claustrophobic if it were not for the very considerable amounts of open land that had been included within the limits of the Greater Berlin of 1920. West Berlin has 40.3 sq. m of forest and 23.3 m of other recreational land for each inhabitant. It is quite remarkable that, within the boundaries of a great city it is possible to walk or ride for hours through fields and forests, travel a whole chain of lakes by yacht, motor cruiser, or pleasure boat, row the measured regatta course in the Havel, swim in summer, skate in winter, or even, in what is supposed to be a plain, ski (on grass in summer, on snow, if need be machine-made, in winter) on the enlarged Teufelsberg.

Berlin's recreational facilities are enormously enhanced by the inheritance of the Ice Age. The great chain of lakes along the Havel is accompanied by moraine and kame features, the highest country in West Berlin, providing quite dramatic variations in elevation. Extensive forests line the lake shores, the greatest of which, the Grunewald, is crossed by a former sub-glacial valley occupied by another chain of lakes. Eskers, for example the Wallberg running east from Tegel, and areas of sand dunes, add further variety. There is even land in continuing agricultural utilization, notably at the isolated locations of Gatow and Kladow, west of the Havel, and to the north at the village of Lübars (Table 3.2).

These natural resources are just a surviving residue, if a most valuable one, of land that has not yet been absorbed by urban expansion. Even public ownership was no defence to the disposal of forest land for other purposes; for example, in the last hundred years

Table 3.2 *Land use in West Berlin, 1950 and 1985*

Form of use	1950 ha	1950 %	1985 ha	1985 %
Built-up areas	13,581	28.2	20,263	42.2
Streets, squares	5,267	11.0	5,942	12.4
Railway land	1,249	2.6	1,254	2.6
Building sites	3,596	7.5	–	–
Parks, playgrounds	1,347	2.8	4,191[a]	8.7[a]
Cemeteries	577	1.2	673	1.4
Agricultural land	9,415	19.6	2,304[b]	4.8[b]
Forest	8,197	17.0	7,695	16.0
Water	3,016	6.3	3,246	6.8
Other (including airfields)	1,850	3.8	2,446	5.1
Total	48,095	100.0	48,014	100.0

Notes:
Sources: Berlin in Zahlen 1951; *Statistisches Jahrbuch Berlin* 1986
(a) Official sources give 10 per cent of the total area as devoted to recreational needs, including (1981) 8.7 devoted to parks, sports grounds, playgrounds and gardens. In 1981 garden colonies occupied 1,910 ha or 4.0 per cent of total area.
(b) Includes rough grazing (heathland and bog); farmland as such accounts for only 2 per cent of total area.

the Grunewald forest has lost great areas on its south-east and north-east fringes for residential development. It has clearly been difficult to resist calls for the allocation of forest land for worthy and often urgent public purposes, such as publicly backed housing projects, sports facilities, schools, hospitals, tips for rubble from the bombed city, and further tips for domestic waste when West Berlin was isolated.

The forests also provide attractive possibilities for the unhindered development of means of communication; for example, the Grunewald has at various times lost ground to the Heerstrasse (the military road to the west), the railway and accompanying S-Bahn to Potsdam, the nearly parallel U-Bahn (here above ground), and the inter-war proto-autobahn known as the Avus, widened to autobahn standards after the Second World War, with additional tree felling to provide space for access points and the Dreilinden control station at the GDR boundary. Forest was also lost for the building of Tegel Airport. More recently, in a more environmentally conscious age, public protests delayed the building of direct access through the Tegel Forest to the new autobahn linking Berlin with Hamburg across the GDR, although by early 1986 it appeared that all legal steps to prevent the road had been exhausted, and that construction by 1988 was likely.

The military have at all times been consumers of forest and

heathland, for training areas or for 'temporary' camps that all too often became permanent. In the National Socialist period a start was even made on a new university, with a 'Military Science Faculty' in the Grunewald, on the site now occupied by the Teufelsberg rubble tip. The last military incursions came with the arrival of the western Allies, when the Americans established barracks and quarters on the fringe of the Grunewald and the French on the Jungfernheide at Tegel Airport. The same is true of the British military establishment in Gatow, which impinged on the Gatower Heide forest, and where in 1986 controversy flared because of a proposal to fell 5,500 trees to meet new safety standards on the approach to the airfield. Parts of the forest are still used by the western Allies for training purposes; a British firing range at Gatow has been particularly controversial, on grounds of noise pollution.

The Berlin forests suffered considerable damage during the war years, culminating in the Battle for Berlin. Then in the immediate post-war years of fuel shortage and blockade thousands of trees were felled; of something approaching 8,000 hectares of forest, 2,000 hectares were clear-felled and a further 3,000 hectares so badly affected as also to need replanting, so that even today West Berlin's forests have a high proportion of young trees. At first there was a return to 'economic' forestry, with the planting of uniform stands of pines, but increasingly emphasis has been placed on amenity value. Especially in areas in the vicinity of water, favoured by visitors, the proportion of deciduous trees is increased and paths developed for walkers, riders, and health enthusiasts. Any available animal life, from ant heaps to deer herds, is carefully preserved and displayed, while forest restaurants meet the appetite for coffee and cakes. The forest area is now vigorously defended; during the three decades, from 1950 to 1981, there was a reduction of from only 8,197 hectares to 7,650 hectares, or from 17 per cent to 15.9 per cent of the total area of the city.

West Berlin's natural environment is almost as much in danger from those who love it as from the builder; pressures from a whole range of recreational and urban-fringe phenomena are understandably intense where availability is so restricted. The passage of so many feet rapidly wears down the forest paths to erodable sand, especially on slopes. Wash from motor boats erodes the reed beds along the banks of lakes and rivers, which are an important environment for amphibians and water birds. The growth of boat yards and marinas has been strictly restricted by the planners, or they would by now have lined every shore; consequently, the scores of thousands of yacht and boat owners have difficulty in finding a mooring and must pay heavily when they are successful. Access to the water is also a problem; in the vicinity of the various lakeside settlements private gardens go right to the water's edge; either there is no path for the public or it runs some metres inland, fenced off from the water by a row of private moorings. This increases pressure on the available access points, especially if served by road, where on sunny weekends traffic and parking pressures can be formidable. The crowding at the limited number of bathing places is undoubtedly one of the most disliked features of this otherwise highly appreciated recreational area, but is a function of the travel difficulties that bar West Berliners from the hundreds of lakes beyond their boundary.

Particularly fragile areas, such as some of the lakes and surviving bogs in the Grunewald sub-glacial valley, can be given legal protection by being declared nature reserves (*Naturschutz*); there is usually no public access. Other areas are preserved for their particular visual value (*Landschaftsschutz*), while particular objects of scientific or historical interest, from a single tree or glacial erratic to areas of special flora are protected as natural monuments (*Denkmalschutz*).

Agricultural land has been the major victim of postwar urban expansion, and what survives is under particular pressure. The use of the land is strongly influenced by a combination of high wage rates and urban demand in an area of restricted land availability. Rather surprisingly, some of the lighter soils are used for growing grain, especially rye, in part because of the limited labour requirements of mechanized cultivation, in part because of the additional income obtainable by the direct sale of straw to Berlin's horse-owners, in part because of high federal subsidies. There is a marked concentration on the growing, where necessary with spray irrigation, of crops directly intended for the Berlin market, in part through direct sale. Characteristic products are fresh vegetables, kitchen herbs, raspberries, and straw-berries; the soft fruits are also marketed through 'pick-your-own' arrangements. Eggs and milk are also produced for roadside sale (Haserodt 1985).

Many of these new activities, such as horticulture, greenhouse cultivation, fruit growing, or intensive egg production, involve con-siderable changes in the traditional agricultural landscape, which is one of open arable fields on the higher areas and grazing land in the depressions. The most remunerative activities are, however, scarcely agricultural at all, and lead to conflicts between local interests, desirous of increasing the return on their land through leisure-related activities, and those wishing to preserve existing agricultural practices as a means of preserving the landscape. The growth of that typical urban-fringe activity, riding, is a particular source of dissension. It is common for farm buildings, even those belonging to agricultural enterprises that are still active, to be adapted to provide stabling, where urban dwellers can keep their horses at livery at vast expense.

It is not a big step from this to the building of indoor riding schools and the division of increasing areas of the former arable fields into paddocks. Controversy over riding developments has particularly centred on the village of Lübars, on the northern fringe of Berlin, which is beginning to look more like Newmarket than a typical village of the Brandenburg Mark (Vollmar 1985). A form of land use providing income from remote and inconvenient corners of the farm is the provision of camping and caravanning sites, another activity likely to produce conflicts with the land planners.

A particular urban-fringe problem is provided by the sewage farms, of which West Berlin has only one, at Gatow, west of the Havel. A part is still kept for emergency use in the unlikely event of a breakdown in the Ruhleben treatment plant, but finding alternative uses for the remainder faces the problem of heavy-metal contamination from long-continued effluent disposal. As an interim measure, part of the area has been given over to the most extraordinarily ramshackle collection of urban-fringe activities of them all, including greyhound racing, dog training, model plane flying and, inevitably, horse-riding establishments.

The West Berlin government is now determined that the erosion of the surviving 'natural' heritage shall cease. Building plans have been amended to limit further development to existing built-up areas, leaving the surviving areas of continuous open space untouched. This is however a policy that gives concern to the Chamber of Commerce and others involved with Berlin industry, who fear that the lack of 'green field' sites will reduce the attractiveness of the city to new industrial projects.

3.7.3 Garden colonies

Characteristic of the great German cities is the attempt to counterbalance their high residential densities by the creation of gardens for individual use on the city fringes. These are the *Kleingärten*, or garden colonies, sometimes known as *Schrebergärten* (after Dr Schreber, a pioneer of the movement in Leipzig) or *Laubenkolonien* (summer-house settlements). Unlike the British allotments, mere partitioned fields where the wind blasts unchecked through the sodden brussels sprouts and tools are kept in shanties made of old packing cases, the Berlin garden colonies are extremely orderly and well-organized affairs, made attractive by their mature fruit trees. Behind formidable fences and locked gates, the profusion of flowers, the patches of lawn, and the garden furniture make their primarily recreational function apparent, while the 'summer-houses' can approach the solidity and dimensions of cottages, in which the family can spend the night. Retired people can even stay out for the whole summer.

In the period of acute housing shortage caused by wartime bombing, many 'summer-houses' came to be permanently occupied, producing a kind of untidy suburbia. Not only was this a planning problem, it was a health problem, as garden areas were often low-lying and were not linked to the sewage system. Transformation into permanent housing areas was one solution, as at the giant Märkisches Viertel housing project in northern Berlin, where some garden colonies still survive among the tower blocks.

As garden colonies are usually established on land in public ownership, they are a tempting target where space has to be found for new housing projects, motor highways or other public projects. It may be thought that providing these low-rent gardens for a few thousand families is an extravagant use of the limited recreational land of a crowded city, but their possessors are quick to spring to the barricades in defence of their individual paradises. Such action ensured, for example, that the 1985 Garden Show and later public park for half a million inhabitants in southern Berlin, would be linked to its neighbouring settlement nuclei by relatively narrow avenues through the garden colonies, not by wide-sweeping lawns. There has, in fact, been a considerable decline in the number of gardens since the end of the Second World War, especially when it is remembered that the many gardens situated on GDR territory became inaccessible to West Berliners from 1952. West Berlin policy is now to keep the number of surviving gardens at approximately the present level, but to group them into sites where the occupants can be given reasonable security of tenure. These would make some contribution to general recreational value by being linked and traversed by the system of green walkways that is under development in the city.

3.7.4 Recreation in urban West Berlin

Social disparities in access to recreational facilities are reinforced by spatial disparities of supply. The Wilhelmian ring, built-up before 1914, has the highest residential densities in the city, the lowest housing standards, and the greatest proportion of low-income people with restricted opportunity to holiday outside the city (see section 8.5.1). Yet it is precisely in these areas that open space is most restricted. As compared with a West Berlin average of 63.2 sq. m of recreational and forest land per inhabitant, the inner-Berlin *Bezirk* Kreuzberg has only 3.6 sq. m, while peripheral Zehlendorf, with its numerous high-income residents, has 343.5 sq. m (Steinecke 1985).

The greatest of the inner-urban open spaces is the Tiergarten, which started its existence as an electoral game preserve but which was changed in successive stages from the seventeenth century onwards into a vast 167-hectare park; at its south-west corner is the 1844 Berlin Zoo. The whole area was devastated in the Second World War, when most of the surviving trees disappeared into the stoves of Berliners in the shortage winter 1945–6. Parts were even parcelled out as allotment gardens for growing vegetables. Rehabilitation began in earnest immediately after the end of the blockade: today the Tiergarten presents an informal landscape of winding paths in a setting of wood, water, and grass. Islands of more formal planting include the rose garden, a post-war gift from Britain. The Tiergarten is traversed by Berlin's formal axis, the Strasse des 17 Juni (formerly Charlottenburger Chaussée). On a summer Sunday it is possible to gauge the significance of this open space for Berlin. Sunbathers strew the lawns, and everywhere close to roads, where parking is possible, the smoke rises from the portable barbecues; the large Turkish family groups particularly catch the eye. Yet the Tiergarten, bordered on the east by the Wall, on the north by the Spree and the little-frequented Reichstag building, and on the south by the ruined embassy quarter, is really rather cut off from most of Berlin; it is particularly inaccessible for the dense-packed inhabitants of Kreuzberg, Tempelhof, and Neukölln in the Wilhelmian ring south of the East Berlin centre.

A new conception of communal responsibility for public health in the mid-nineteenth century led to the creation of a few small parks in the congested inner areas, typically sited on, or close to, the edges of the till plateaus, overlooking the *Urstromtal* (see fig. 6.2). To the north, Volkspark Humboldthain contained a memorial to the famous geographer, destroyed in the Second World War and replaced in more simple form. The park also contains one of the rubble tips created after the Second World War, enclosing a bunker, the top of which gives a splendid view of the city. Situated amid the industrial plants and dense-packed housing of Wedding, the park is much appreciated by the population. South of the *Urstromtal*, Viktoriapark in Kreuzberg also has splendid views over the inner city. Neighbouring Volkspark Hasenheide in Neukölln has another rubble mountain, bearing a memorial to the *Trümmerfrauen*, who laboured to clear the war-ravaged city of its ruins. Of interest also is the small Kleistpark, a stage in the 1679–1896 outward migration of the Berlin Botanical Garden from the Palace to its present location in Steglitz (see fig. 6.3). It currently contains the 1913 Kammergericht, a building now devoted to the largely functionless Allied Control Council, before which the flags of the former Allies of 1945 meaninglessly fly.

To supplement the meagre provision of open space in the inner parts of West Berlin, it has been consistent policy in rebuilding after bombing, and in the urban renewal programmes for the older housing stock (*Mietskasernen*), to provide small areas for children's playgrounds, for ball games and for older people to sit out. It now seems possible that the land occupied by the derelict railway termini and their approaches along the south side of the former Berlin urban core can be transformed into parks, as well as some of the former embassy quarter south of the Tiergarten, which cannot be built on because it belongs to foreign states.

In the inter-war period the Weimar administration added further people's parks on the fringe of the nineteenth-century close-built residential area. These included Volkspark Jungfernheide in Charlottenburg and Volkspark Mariendorf in Tempelhof. After the Second World War, the West Berlin administration continued to endeavour to provide additional public open space, in spite of the increase in the built-up area of the city from 34.4 per cent in 1965 to 40.3 per cent in 1981 (Steinecke 1985). New parks were created, for example at Lübars, in the extreme north, to cater for the inhabitants of the peripheral Märkische Viertel housing project, or on the site of the 1985 Garden Show, to provide for the people of the south-east of Berlin. One of the most interesting developments is the attempt to link all the fragmented open spaces in the city by walkways, following in part waterways such as the Spree, the Landwehr Canal, or the Teltow Canal.

3.8 Environment and recreation in East Berlin and its region

3.8.1 Open spaces within East Berlin

The hazard of division of Berlin left East Berlin without a major central urban space on the scale of West Berlin's Tiergarten, although the extremely spacious layout of the new East Berlin centre has provided a widespread arrangement of lawns and gardens.

Within the Wilhelmian pre-1914 area of high-density development, Berlin GDR has one of the earliest nineteenth-century parks on the Barnim fringe, Volkspark Friedrichshain, overlooking the *Urstromtal*. Its landscaped rubble tips have splendid views over the Berlin inner city. Redevelopment of the high-density *Mietskasernen* area has, as in West Berlin, provided small play areas and gardens for elderly persons to enjoy. An outstanding development has been the transformation of the former Berlin gasworks in Prenzlauer Berg into a housing area with extensive gardens surrounding the monument to Ernst Thälmann.

The communal ideology of the German Democratic Republic means that every effort is made to provide public recreational facilities for young and old. A number of these are found on either side of the Spree, upstream from the densely built-up area of the city. They include the new zoo at Berlin-Friedrichsfelde, the Pioneer Park especially for young people at Oberschöneweide or the complex of parks at Treptow, which includes the massive Soviet War Memorial, an 'Insel der Jugend' (Youth Island) in the Spree and a 'Kulturpark' (Amusement Gardens). The large-scale housing projects on the east and north-east fringe of the city are also amply provided with public gardens and children's play areas, and new parks and sports areas serving them are being developed.

3.8.2 The East Berlin rural fringe

In terms of recreational possibilities, West Berlin is mirrored by East Berlin, with the assemblage on its eastern fringe of lakes such as the Müggelsee, set in great forests, with hills rising to the maximum for Berlin of 115 m (83 m above the level of the lake). This area, contained with the East Berlin *Bezirk* Köpenick, is well served by S-Bahn, tram, and bus, and is the principal target of journeys for open-air recreation within the city. The available facilities generally repeat the pattern already described in the Havel area of West Berlin, although with a stronger element of public enterprise. The Müggelberge are crowned by the Müggelturm, a tower reaching 122 m above sea-level, constructed in 1960–1, and giving widespread views over the whole lake area. As it can be reached by road, and has a large restaurant, it is much frequented.

The proportion of journeys for short-period recreation contained within the city is, however, dropping steadily: the figure was 53 per cent in 1970, 41 per cent in 1977 (Seibicke 1983). The East Berliner is not confined to his city and can seek out less-crowded lakes, hills, and forests well beyond its boundary; there are said to be 3,000 lakes within a radius of 150 km from the city, although most short-period recreation takes place within about 50 km, or a ninety-minute journey by public transport. The increasing proportion of journeys taken outside the city undoubtedly reflects the increasing motorization of the East Berlin population. The establishment of large new housing projects close to the east and north-east boundary of the city is another factor; from these the adjacent countryside is easily reached by individual motor transport.

East Berlin and its surrounding area has a parallel system of nature reserves (*Naturschutzgebiete*) and areas of protected landscape (*Land-schaftsschutzgebiete*) to that in West Berlin. In addition, a number of more extensive areas of forest and lake are specially earmarked for short-term recreation. These *Naherholungsgebiete* (NEG) are distributed in a half-circle around the eastern exterior of the city; the additional time involved in circumventing the obstacle of West Berlin makes the area west of the city unfavourable for short-period recreation, although it contains some magnificent countryside, notably the

Potsdam lakes. The NEGs benefit from special measures designed to keep open the access to lake and river shores, to preserve water purity and fish stocks, and to provide an appropriate array of services.

While many recreational journeys into the surrounding countryside are purely day trips to some lakeside or forest, there are also longer-term possibilities. Camping is very popular, not only with Berliners but with visitors from the rest of the GDR and even from other countries. The sites are mostly strung out along the lakes south and east of the city. There are also numerous holiday homes set up by trade unions and individual enterprises, camps to give holidays to school children from the city, and youth hostels (Spitzer 1975).

The most original form of short-term recreation is, however, provided by the swarms of lightly constructed individual summer homes, often referred to in the literature as bungalows, but commonly called dachas. The dachas have a measure of official toleration, verging at times on approval; after all, the members of the ruling politburo have their own exclusive dacha settlement outside Berlin. At least at times, it has been possible to purchase wooden-frame buildings for self-erection, and to find a local council prepared to allocate a stretch of land to be divided into lots for dacha construction. They are, however, intended for temporary occupance only, usually having neither piped water nor main drainage.

The attraction of a dacha to a family spending the working week in a small Berlin apartment is obvious, offering the possibility of active and passive leisure in rural surroundings outside the city. To some extent also the dacha replaces the holidays abroad in the GDR that are difficult to arrange on an individual basis. Could it also be that they offer an escape into a private world from the very strong pressures to adopt approved forms of behaviour that are characteristic of the GDR? Certainly, the possessor of a dacha tends to be passionately involved, spending hour upon summer hour working in its garden, and adding yet one more extension to the original simple core. This process of accretion means that dachas are inclined gradually to turn into suburban houses, a development unwelcome to local administrations, who had never expected to be called on to provide an urban level of services. Altogether the dachas are a surprising phenomenon in a country otherwise devoted to the state provision of virtually everything.

4.1 Long-distance transport

The evolution of Berlin from the mid-nineteenth century onwards into a major traffic node by road, rail, water, and air has already been mentioned (see sections 1.2 and 1.4). Except to some extent in relation to waterways, this nodal position was not ordained by nature, as might be said of Magdeburg or Dresden. The most that can be said is that the terrain of the Northern Lowland offered no serious obstacle to transport developments, which were essentially related to the rise of Berlin to capital status. But what was given by human action can be taken away; the fateful post-war reversal in Berlin's political relationships had inevitable and severe consequences for its transport relationships. The 'island' of West Berlin, with the loss of its capital functions, has also lost much of its weight as an originating point and destination for flows of goods, of information, and of people. It is also no longer a place of exchange; people no longer travel to Berlin to change on to another means of communication for onward travel. Much the same is true of goods; Berlin is simply the point of origin or point of destination, no longer an intermediary for more widespread trade. This also brings up the problem of return freight; West Berlin is a consumer of bulk commodities, but the vehicles that carry them must generally make their return journey empty. West Berlin has also been shorn of its former close web of communications with the immediate hinterland; instead virtually all outside links have to make the long journey across GDR territory in order to link with the 'mainland' of the Federal Republic.

By contrast East Berlin has a more normal relationship with its hinterland and the neighbouring countries to north, east, and south. It has nevertheless had to make major adjustments to its own lines of communication so as to avoid as far as possible the sealed-off 'foreign body' of West Berlin.

West Berlin's links with the Federal Republic were highly vulnerable, as the 1948–9 Berlin blockade showed. Even after the lifting of the blockade, the land links were subjected to routine frustrations and delays. On the autobahn in particular, travellers had to fill up application forms, purchase a transit visa, pay a special tax for use of the road, and endure searches of vehicles and baggage. All these transactions were conducted in a depressing atmosphere of official frostiness. It was easy for the GDR to find reasons to interrupt traffic, particularly when some political event in West Berlin was causing displeasure.

The efforts of the government of the Federal Republic to mitigate the impact of Berlin's isolation have already been described (see section 2.3.1). The great turn-round came with the 1971 Berlin Agreement. In December of that year, just a few months after the representatives of the four occupying powers had signed the main agreement, the governments of the Federal Republic and of the GDR concluded a further agreement on transit arrangements for civilians and goods between the Federal Republic and West Berlin. This cut out most of the time-wasting frustration of the previous arrangements. The GDR financial exactions were also 'bought out' by the Federal Republic, which has also from time to time contributed money for the upgrading of the various routes. While these new arrangements have made the passage of goods and people a good deal less unpleasant and have largely removed arbitrary and unnecessary delays, the basic facts of West Berlin's isolation and distance from the Federal Republic have not been removed.

4.1.1 Waterways

Of all the means of communication serving Berlin, waterways are the most dependent on natural conditions. The Spree–Havel route is part of the history of Berlin, and is still the key element in the city's water-transport system. The numerous lakes of the city fringes, the pattern of *Urstromtäler* and the low watersheds between streams have all facilitated the creation of a network of waterways surrounding the city. Effective canal building began in the eighteenth century, taking the form of a number of alternative east–west links between the Havel–Spree system and the Elbe to the west and the Oder to the east. These new waterways allowed trade to be concentrated in the capital, to the detriment of towns such as Frankfurt on Oder, which were better equipped by nature.

The subsequent history of Berlin canal-building has been dominated by a series of efforts to supplement the Spree route through the heart of the city, where in the early nineteenth century delays of up to a week could occur at the locks. As a general rule, it can be said that the improvements never quite caught up with the advancing size of the available carriers, so that there have been continuing complaints about locks that are too small, and about capacity limitations because of low water in dry periods. Even today the system will not take the standard European 1,350-tonne canal barge.

A first southern bypass was provided by the building in 1845–50 of the Landwehr Canal, outside the line of the Customs Wall; today its green-planted banks are a welcome addition to the urban scene. In 1848 work began on the 'Spandauer Schiffahrtskanal', designed to provide a northern bypass from the Tegeler See to the Nordhafen, a route that was improved in the early twentieth century by the building of the Hohenzollern Canal. Plans which had been discussed since the middle of the nineteenth century for an additional outer bypass canal came to fruition at the end of the century with the construction of the 'Kanal des Kreises Teltow' (Teltow Canal). With its numerous harbours the canal was conceived from the beginning as a means of attracting industry to new sites in the Berlin Outer Zone (see section 5.3). A number of short canal spurs from the Havel in the Spandau area were also specifically designed to provide industrial sites. The Westhafen Canal was added in 1954–6 to avoid the locks at Spandau, then too small to take 1,000-tonne vessels (fig. 4.1).

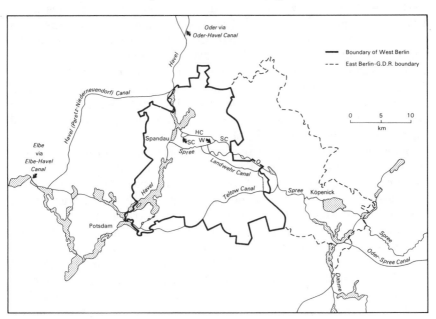

Figure 4.1 Berlin's waterways. SC: Spandau Canal; HC: Hohenzollern Canal; W: Westhafen; O: Osthafen

There is no single canal harbour for Berlin but instead a number of harbours dispersed around the system, as well as a large number of quays belonging to individual firms or public utilities. The nearest approach to a main harbour is provided by the Westhafen, finally completed in 1927. The harbour consists of three parallel basins, whereas the 1913 Osthafen, the second most important habour of pre-war Berlin, was formed by riverside quays at Stralauer Tor (Oberbaum), upstream from the core of the city.

The division of Germany and the isolation of West Berlin affected all means of transport, but waterways were something of a special case, since their dependence on natural features means that they are not easily moved or readily abandoned. Given that West Berlin occupies a nodal position astride the waterway system of the GDR, both East and West have a common interest in keeping the waterways open to traffic.

A partial waterway bypass of West Berlin was provided by the building of the Paretz–Niederneuendorf canal, allowing Havel traffic to avoid the city, but GDR traffic following the Spree still needs to pass along one of the West Berlin waterway routes. The Teltow Canal was blocked at its south-west end, presumably because it was mainly of interest to West Berlin industry rather than for through traffic. West Berlin traffic had to make a tedious detour through the heart of Berlin and the GDR control point there in order to enter the canal from its eastern end, adding two days and additional monetary exactions to the round trip. After thirty-six years the south-western access to the canal was reopened following the 1978 transit agreement between the Federal Republic and the GDR. Negotiations were protracted by the now standard GDR attempt to deal directly with the senate of West Berlin, thus implying that the latter has some special 'free-city' political status independent of the Federal Republic, which claims (and has hitherto maintained) that it is to be the sole negotiator of agreements at government level.

Shipping interests in West Berlin also complain about periodic, often unexplained and probably discriminatory restrictions, such as the banning of propulsion units of over 600 horse power (which the West Germans tend to rely on but not the East Germans), restrictions on routes that may be used in, for example, transit trade with Poland, and limitations on the permitted size of vessels. To the extent that the restrictions reflect the genuine deterioration of the waterway system, they can to some extent be 'bought out' by payments from the Federal Republic for dredging and repairs. As part of its general assistance to the Berlin economy, the federal government also 'buys out' the fees that would otherwise be imposed on transit trade at the GDR control points.

A particular difficulty is the condition of the Mittelland Canal, designed to be the principal east–west waterway link in Central Europe. This was complete from the Rhine to Magdeburg when work was stopped by the outbreak of the Second World War. The section in the Federal Republic has been enlarged to take 1,350-tonne 'European' barges. At Magdeburg, however, Berlin-bound barges must descend by lift to the Elbe, which has to be followed for 11 km to the entrance of the Elbe–Havel canal leading to Berlin. The Elbe is a river of notoriously erratic regime, and in the low-water period between May and September barges can often use it only if partially laden, thus increasing transport costs. Work had begun at the outbreak of war on a canal bridge over the Elbe and on a new barge lift to give direct connection to the Elbe–Havel canal. Completion of these projects would benefit traffic not only between Berlin and the Rhine but between Berlin and Hamburg; obviously the federal government would be required to meet most of the bill. A longer-term dream is the improvement to 1,350-tonne standard of the entire waterway between Elbe and Oder, as part of a grand Black Sea–Dnepr–Pripyat–Bug–Vistula–Noteć–Warta–Oder–Berlin–Havel–Mittelland Canal–Rhine route.

A further problem lies within Berlin itself, caused by the inadequate size of the Spandau lock on the Havel. This not only obliges the larger West German boats to make a detour through the heart of West Berlin to reach the upper Havel, it also inconveniences the GDR push-units, which have to be broken up to pass the lock. It might be thought that the building of a second lock would be of such obvious benefit to all that it would be put in hand quickly once finance was assured, but the technical operation of locks in the British sector of Berlin had in 1951 been entrusted by the British commandant to the *Wasserstrassen-hauptamt* (waterways administration) in East Berlin, and there were the usual delays caused by an attempt to make approval dependent on completion of a 'state treaty' between the GDR and 'independent' West Berlin. The project was eventually approved, only to run into opposition from West Berlin environmental groups, fearful of the effect of the passage of larger vessels upon the foundations of the Spandau Citadel.

Waterways carry about 30 per cent (by weight) of the commodities needed to sustain West Berlin, less than moves by road, but more than rail or air. The balance between quantities received and despatched is highly uneven: 3 million tons of goods are transported from the Federal Republic to West Berlin, but only about half a million tons move in the reverse direction. Understandably, water transport is mainly used for bulk commodities, above all petroleum products, but also coal, bulk building materials, non-ferrous metals, and iron and steel products.

4.1.2 Railways

In the forty years that followed the opening of the Berlin–Potsdam rail line in 1838, Berlin developed into the centre of a spider's web of eleven radiating rail lines. There was no single main station (*Hauptbahnhof*); instead, just as in London or Paris, the lines ran separately into individual terminal stations. These were situated on the fringe of the built-up area as it stood at about mid-century, that is, very broadly, along the line of the Customs Wall (fig. 4.2). The stations were, in anti-clockwise sequence, the Lehrter Bahnhof in the north-west (replacing the 1846 Hamburger Bahnhof) for lines to Hanover and Hamburg, the Potsdamer Bahnhof, the Anhalter Bahnhof (the greatest of them all, starting point of important lines to industrial Saxony and beyond into Central and Southern Europe), the Görlitzer Bahnhof, the Frankfurter (later Schlesischer, later Ost) Bahnhof, and the Stettiner (later Nord) Bahnhof. There were repeated proposals for the creation of a single main station, or at least for the concentration of lines into only two or three stations, of which the Hitler/Speer proposal for a vast new south station near Tempelhof Central Airport was only one. Nothing was achieved before the post-war division of the city imposed more radical solutions. The building in the late 1860s of the S-Bahn ring line, followed by the east–west S-Bahn 'diameter' line through the heart of the city, considerably eased the problem of interconnection between the terminal stations.

Closely associated with the terminal stations, the great goods stations also lay close to the core of the city. At a time when road transport was still horse-drawn, it was essential that goods stations should be in proximity to the ultimate destination of consignments. The passenger stations, their approach lines, and the goods yards provided major interruptions to the continuity of the urban fabric, especially where a number of lines ran in parallel, as in the approaches to the Potsdam and Anhalt stations. Goods stations and their associated marshalling yards were a particular source of noise and traffic. An opportunity to move at least the marshalling yards out of the urban area was provided by the beginning in 1912 of an outer rail ring for Berlin goods traffic. This project was completed only after the Second World War, in response to the GDR's wish to avoid the necessity for its rail traffic to pass through West Berlin.

The post-war occupying powers did not consider it practicable to control and operate rail services in the restricted area of West Berlin separately from the system in East Berlin and the GDR. Accordingly, inter-Allied agreements left the whole of the Berlin long-distance rail services (as well as the S-Bahn) in the hands of the GDR-based Deutsche Reichsbahn. The great terminal stations have for the most

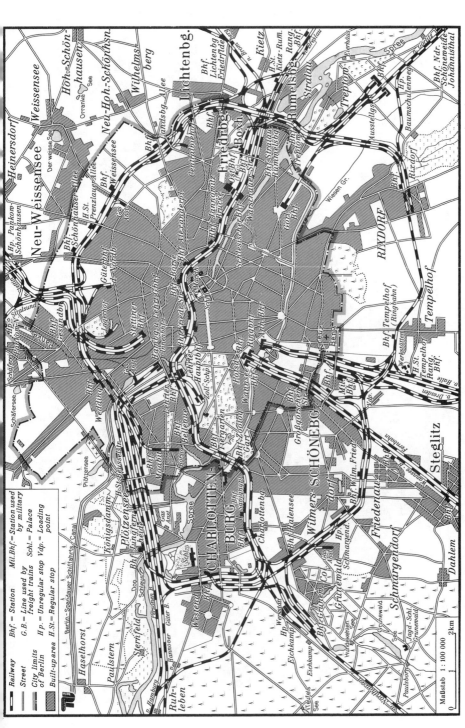

Figure 4.2 Railway installations inside the S-Bahn ring (Source: Hofmeister 1985)

Plate 4.1 The remains of Anhalt rail terminus, 1986. The space formerly occupied by the platforms and approach lines, long a lorry park, was, after use for an exhibition connected with the 750th anniversary celebrations in 1987, expected to be

part been abandoned. The Potsdam station site is invisible somewhere beneath the installations of the Berlin Wall, the Anhalt station a fragment of façade giving on to an open space where the platforms once stretched. The Lehrte and Görlitz stations are similarly abandoned. It is somewhat ironic that the Hamburg station building, the oldest of them all and turned into a transport museum, has survived relatively unscathed. The approaches to the Potsdam station formed a curious tongue of *Bezirk* Mitte and thus of East Berlin at its extreme south-west angle, which was a distinct obstacle to West Berlin road traffic. It was transferred to West Berlin as part of a set of exchanges of such boundary anomalies. West Berlin has also purchased the disused railway land from the Deutsche Reichsbahn; it was for many years used for a variety of 'fringe' uses (see section 7.4), but its long-term use is probably for parks and sports facilities.

So far as long-distance passenger traffic is concerned, West Berlin was restricted to the former S-Bahn 'diameter' line, stopping only at the Berlin Zoologischer Garten station before entering East Berlin at the Friedrichstrasse crossing point. This can be regarded as the pale survivor of the once great European line linking Channel ports, Paris, Berlin, Warsaw, and Moscow. A direct link to the Hamburg line was opened later.

Rail traffic between the Federal Republic and West Berlin is restricted to a limited number of transit lines, on which only a specified number of trains per day may run. Passenger trains are restricted to four corridors and interzonal crossing points: Schwanheide–Büchen (for Hamburg), Marienborn–Helmstedt (for Hanover), Gerstungen–Bebra for Frankfurt, and Probstzella–Ludwigstadt for Munich (to which, since 1972, Gutenfürst–Hof has been added, giving a shorter route). All passenger trains originating in West Berlin, together with all military trains of the western Allies, irrespective of destination, start off in a westerly direction. Until 1976 all had to pass the single GDR control point at Griebnitzsee, only thereafter assuming their appropriate directions towards Hamburg, Hanover, Frankfurt or Munich, a procedure inevitably involving some loss of time. Following the 1975 Transit Agreement between the governments of the Federal Republic and the GDR, an additional rail crossing point was opened in 1977 at Staaken on the Hamburg line, reducing the journey time by more than half an hour. In addition, the stations of Charlottenburg, Wannsee (on the Griebnitzsee line), and Spandau (on the Hamburg line) were opened to international traffic, so that people living in the west and south of West Berlin no longer had to travel to a single long-distance station at Berlin Zoologischer Garten. Passenger trains to and from the Federal Republic are not available for internal GDR passenger traffic. Initially, goods traffic between the Federal Republic and West

Berlin was permitted to use only the Helmstedt–Marienborn interzonal crossing, but since 1965 the other crossing points mentioned above have been permitted (fig. 4.3).

The government of the Federal Republic regularly pays sums of money to the GDR government for improvements on the transit lines

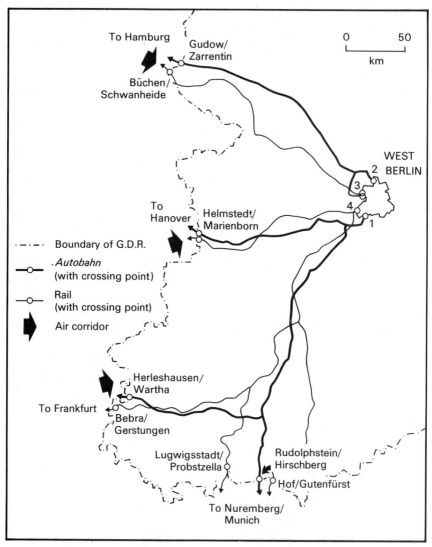

Figure 4.3 The West Berlin transit routes. Berlin crossing points: 1: Dreilinden/Drewitz (Autobahn Hanover, Frankfurt, Munich); 2: Heiligensee (permanent crossing for Autobahn Hamburg); 3: Staaken (Hamburg rail and provisional Autobahn approach); 4: Wannsee/Griebnitzsee (rail Hanover, Frankfurt, Munich)

and within West Berlin; in late 1986 consideration was being given to contributing to the total rebuilding and electrification of the Helmstedt–Berlin line, since 1945 reduced to a single track, the cost of which was estimated to involve a larger sum than the total contributed to the GDR for all forms of transport up to that date. Completion of the project would enable the other crossing points to be abandoned; passengers from West Berlin would use the new line to Hanover, before changing to the fast inter-city links to all parts of the Federal Republic.

The long-distance rail system of East Berlin is something of a mirror-image to that of West Berlin, although without the same political restrictions. The Ostbahnhof is the principal originator of passenger trains, supplemented for some services by Lichtenberg and Karlshorst stations. The key to the long-distance railway system of East Berlin is the completed outer rail ring round Berlin. All services start in an easterly direction, then divert if necessary onto the outer ring, before diverging, as appropriate, in the direction of Leipzig or Magdeburg.

The outer goods ring around Berlin was still only fragmentary after the Second World War. In the 1950s it was hurriedly completed to allow the avoidance of West Berlin, sometimes with only a single track. The ring line has been under constant improvement ever since (with consequent traffic dislocation) to bring it up to four tracks, thus allowing the separation of long-distance and local traffic. Originally, it had been assumed that while goods traffic would use the outer ring, passenger trains would, where appropriate, pass through West Berlin, for example to Potsdam and beyond. With the building of the Berlin Wall in 1961 this was no longer possible, and passenger traffic had also to be diverted to the outer ring, causing a great deal of confusion and congestion. Difficulties were particularly acute along the southern section of the ring as far as Potsdam. This has for years been chronically affected by works connected with track improvements and electrification. All too often passengers have been required to leave their train to wait in mud and rain before fighting their way onto buses to resume their journey. It is not surprising that Potsdam, once very firmly linked into the Berlin urban system, has thanks to its new isolation been more and more able to develop as a geographically independent capital of a GDR *Bezirk*.

The Ostbahnhof, the principal East Berlin station for long-distance traffic, was extensively rebuilt in preparation for the city's 750th-anniversary celebrations. It is apparently to be called in future 'Hauptbahnhof Berlin'. This change brought fears in West Berlin that the international services through the station at Berlin Zoologischer Garten would at some stage in the future be diverted round the completed outer ring to the new station, and that West Berlin would become a dead-end, with long-distance trains terminating at Charlottenburg.

4.1.3 Road links

When West Berliners think about the transit links with the Federal Republic, it is almost certainly the road (or more strictly autobahn) links that will come first to mind. Not only do the roads bring more goods to Berlin than any other means of transport, they also carry the largest number of travellers. It is as occupants of private cars that West Berliners are most likely to experience the transit arrangements and to come into face-to-face contact with GDR officialdom.

The 1913–21 Avus (*Automobil-Verkehrs- und Übungsstrasse*) through the Grunewald was linked in 1940 to the German autobahn system initiated in the National Socialist period. Reflecting Berlin's capital function, a ring autobahn was planned surrounding the city to act as the point of origin of a system of radiating routes, but it was not completed until 1979. With the division of the city the Avus was for many years the sole West Berlin access to the autobahn system, through the control point at Dreilinden. Crossings onto the ordinary road system were created in the extreme south-east of West Berlin to give access to Schönefeld Airfield, and at Staaken west of Spandau for local traffic and transit traffic to Scandinavia. There are, of course, also a number of crossings into East Berlin. In contrast to the scarcity of links from West Berlin to the autobahn system, all the main East Berlin radials connect with the ring; the links from Pankow northwards and from Grünau southwards are themselves of autobahn standard.

Transit traffic to and from West Berlin is restricted to a limited number of routes (see fig. 4.3). The shortest and most used route runs due west towards Hanover, crossing the GDR boundary at Marienborn/Helmstedt. Two southerly routes share the same E6 autobahn to the neighbourhood of Jena, where one continues southwards to the Rudolphstein/Hirschberg crossing in the direction of Munich, and the other diverges westwards to the Herleshausen/Wartha crossing and Frankfurt. The final link, to Hamburg, initially used ordinary main roads, but in 1982 was replaced by an autobahn, for the most part financed by the Federal Republic. Access to this was to be given by a new West Berlin autobahn through the Tegel Forest to a control point at Heiligensee, but completion of the road was delayed for some years by political and legal action undertaken by West Berlin environmentalist groups. As an interim measure the planned closure of the Staaken crossing was deferred.

As with other means of surface transport, the Federal Republic makes an annual payment to the GDR in order to 'buy out' the charges that would otherwise be levied on road transport to and from West Berlin. It also contributes large sums towards such improvements as the construction of the Berlin–Hamburg autobahn and the reconstruction of the routes to Hanover and Munich. The Federal Ministry for Inner-German Relations gives very specific instructions to motorists on their way to or from Berlin. They must not pick up hitch-hikers, they must not deviate from the permitted route, they must not leave or even throw away 'western' publications at service areas, they must not exceed the autobahn speed limit of 100 km/hr (fining speeding western motorists is a useful addition to GDR hard-currency earnings), and they must not drive with *any* alcohol in the blood.

4.1.4 Air transport

The former military training ground at Tempelhof was opened in 1923 as 'Zentralflughafen Tempelhof', a new airport for Berlin situated only 5 km from the centre of the city and 6.5 km from the West Berlin Centre, a location which at the time was regarded as an advantage. In 1926 a fusion of two earlier companies concerned with civil aviation formed Deutsche Lufthansa AG (until 1934 known as Luft Hansa), with Tempelhof as its main base. The monumental terminal building of 1936–9 is a sign of the importance attached by the National Socialist government to the new form of transport. In the peak pre-war year of 1938, Tempelhof recorded nearly 64,000 take-offs and handled 247,000 passengers. It was by far the most important airport in Central Europe, with regular flights to twenty-seven inland and forty-three foreign destinations (Mayr 1985).

After the Second World War German participation in aviation was initially forbidden. Only with difficulty did the western Allies obtain and maintain the right to use three air corridors connecting with their troops in Berlin, running from Hamburg, Hanover, and Frankfurt. The 1946 Air Safety Agreement, finally accepting the corridors and setting up the Four-Power Berlin Air Safety Centre, is a rare example of lasting co-operation between the wartime Allies. The vital significance of air access was demonstrated during the 1948–9 blockade. Use of the corridors was however restricted to the aircraft of the western Allies or companies nominated by them. This is the origin of the curious anomaly that Federal Germany's national airline, Lufthansa, does not serve Berlin, which remains a protected market for US, UK, and French companies.

It was possible to equip each of the Berlin occupying powers with an airfield. Tempelhof was in the US sector, the United Kingdom took over the military field at Gatow (its full use involved an exchange of territory with the then Soviet zone) and the French were allocated

Tegel, where a new hard runway was created by heroic efforts during the blockade. The Soviet Union, with fewer spatial constraints, used the field belonging to the former Henschel aircraft firm at Schönefeld, just outside the city boundary. After the blockade, West Berlin civil traffic was initially concentrated at Tempelhof. This field, embedded in a densely populated quarter and surrounded by main roads and other transport routes, was by the end of the 1960s incapable of expansion to provide the long runways required by the increasing size of planes. With the growing frequency of traffic, its environmental disadvantages as a generator of noise and atmospheric pollution and also its air-safety problems became ever more apparent. By that stage also, passenger movements were exceeding 5 million annually, so that handling facilities were overstretched.

From about 1960 more and more flights were transferred to Tegel, and it was decided to construct there the current passenger terminal, opened in 1974. All long-distance flights were concentrated at Tegel from 1975. Because the construction of the new terminal was overtaken by the 1971 Berlin Agreement and the subsequent transit agreements between the Federal Republic and the GDR (which made land transit more acceptable), Tegel has never achieved its designed traffic. The airport has the advantage of being only 8 km from the West Berlin centre, with direct access from the urban autobahn. In spite of its proximity to the centre, it is very largely surrounded by open land, industrial areas, and French military quarters, so that it can be regarded as less environmentally damaging than Tempelhof.

Table 4.1 *Land and air passenger movements to and from West Berlin*

Year	Passengers (millions)		Federal subsidy to air passengers (DM millions)
	Land	Air	
1972	10.4	5.5	70.8
1975	14.2	3.9	42.8
1978	16.7	4.0	40.3
1981	17.5	4.4	97.3
1985	23.7	4.6	–

Sources: Der Tagesspiegel (Berlin), 6 July 1982; *Statistisches Jahrbuch Berlin*, 1972–85.

The 'DDR-Zentralflughafen Schönefeld' is the corresponding main airport for East Berlin; unlike Tegel, however, it serves not only the city but its hinterland and indeed the whole of the GDR. It is favourably located 18 km from the East Berlin centre, in open country on the fringe of the built-up area. It has the great advantage of proximity to the outer rail ring south of Berlin. Main-line trains from the west and south of the GDR use the ring stop at the airport station, and many local trains from Potsdam terminate there. From the other direction, Schönefeld has a direct S-Bahn link with East Berlin. This rail accessibility is important because it enables the GDR as a relatively small country to use the railways rather than internal airlines for feeder purposes. A dual-carriageway road links at Grünau with the access to the autobahn ring and with the city's road system. A boundary crossing point directly into West Berlin gives the possibility of attracting further traffic.

The two airfields have greatly contrasting traffic patterns. Where pre-war Tempelhof had rapidly become the great airline interchange point for Central Europe, Tegel is a dead end. Passengers come to it who need to travel to Berlin, or they leave from it, predominantly for the Federal Republic, or to a lesser extent to destinations in Western Europe or North America (in all, twenty-three destinations in the summer of 1984). Because of the limitations imposed by the air corridors, passengers do not fly into Tegel to change for other European destinations.

Schönefeld by contrast is the main base of the GDR international airline, Interflug. In common with the various national airlines concerned, its main links are with the Soviet Union and the Soviet-style socialist countries of Eastern Europe. Longer links have been established with various non-European destinations regarded as falling, to a greater or lesser extent, into the 'socialist camp', such as Vietnam, Syria, Iraq, Ethiopia, Mozambique, or Cuba, with stops at intervening destinations. In recent years much effort has been put into arrangements for joint services with a whole range of western and neutral countries, including Austria, the Scandinavian countries, and the Netherlands.

The necessity of avoiding Federal German territory imposes some very curious routes on flights to and from Schönefeld. Flights for Amsterdam and Brussels, for example, first fly east to Fürstenwalde, almost on the Polish border, then north until over the Baltic, then across Denmark south of Copenhagen, then south over the North Sea to their destination. The present writer on his first Brussels–Schönefeld flight could not understand why the hostess demonstrated the use of the life-jacket before what, in all innocence, he believed to be a straightforward eastern flight over land!

The ability to offer services to such a wide range of destinations from Schönefeld is in marked contrast to the one-sidedness of Tegel. On the other hand, it has to be accepted that the inhabitants of the GDR are not frequent travellers to foreign countries, at least outside Eastern Europe. The frequency of services to any particular destination tends to be lower for Schönefeld, and the total number of passengers only about half; in 1981 Schönefeld handled 2.1 million, Tegel 4.4 million passengers (Mayr 1985). This is to some extent reflected in the appearance of the two airport terminals. Tegel gives every appearance of being a technologically advanced great-city terminal in search of a role, Schönefeld of being a provincial airport that has somehow stumbled into international significance.

4.2 Urban transportation

4.2.1 Urban transportation before 1961

Until the final division of the city in 1961, the backbone of public transport throughout Berlin was the S-Bahn, an abbreviation variously interpreted as meaning 'Schnellbahn' (rapid transport railway) or 'Stadtbahn' (urban railway system). The S-Bahn has its origins in the notion of a 'Ringbahn', originally derived from the military, who wanted to ensure that in the event of mobilization the various Berlin terminal stations would be linked together by a line that would also serve the Tempelhof training grounds (see fig. 4.2).

The actual position of the ring line was a compromise between the desire to maximize utilization by being as close as possible to the core of the city and the desire to minimize land-acquisition costs by avoiding areas of existing urban development. It must also be remembered that the ring line was in the main designed to improve the transport of goods to and from the industries that were already gathering in inner locations such as Wedding and Moabit; the predominance of passenger transport came later. On environmental grounds the engineers were asked to run the line in a cutting so far as possible; this could be done in the fringes of the Barnim and Teltow plateaus to north and south, but in crossing the *Urstromtal* to east and west the line had to run on embankments or arches, which necessarily acted as barriers within the fabric of the city. Certainly, the wide loop to the west, well away from the major rail terminals, seems to have been motivated by the need to avoid the Tiergarten.

The first part of the Ringbahn was opened in 1871, and the remainder in 1877. After electrification in 1924–9 the journey time for a full circuit of the S-Bahn ring was 63 minutes (Hofmeister 1975a: 227–36). Military considerations after the war of 1870–1 were also behind the building of the 'diameter' Stadtbahn line across the heart of the city, linking the Schlesischer (Ost) Bahnhof, Alexanderplatz, Friedrichstrasse, the Lehrter Bahnhof, the Zoologischer Garten, and Charlottenburg. It used the emplacement of the dismantled fortifications to pass round the northern side of the Berlin inner city and was opened in 1882.

By the interwar years the S-Bahn had thrust radiating tentacles through Berlin's outer zone, where it often determined the location of housing projects, and beyond into the zone of commuter settlements and satellite towns surrounding the city (see section 6.4). A final element was provided at the end of the 1930s by another link across the S-Bahn ring, this time running north–south in a tunnel beneath the heart of the city, intersecting the 'diameter' S-bahn line at Friedrichstrasse station, which developed into a major multi-level interchange of both S-Bahn and U-Bahn lines (fig. 4.4).

The dominant form of street transport in the nineteenth century was the tram, horse-drawn from 1865 and powered by electricity from about the turn of the century. Motor buses were added from 1905. In spite of the existence in Berlin of such a powerful electrical-equipment industry, the creation of an electric underground railway system came late. This seems in part to have related to the difficult terrain conditions, especially in the *Urstromtal* (Leyden 1933: 131). In 1939 the U-Bahn was still a relatively small system supplementing the S-Bahn in the more densely settled inner part of the city.

When the western Allies arrived to take up their Berlin sectors of occupation in 1945, they were persuaded that because the S-Bahn was a unified system crossing the boundaries of the various sectors of occupation and even of Greater Berlin, and because it was intermeshed with the normal long-distance railway system, it should be operated under the single control of a directorate of the former Deutsche Reichsbahn (German State Railways) located in East Berlin. Even the breakdown of the unified city administration in 1948 left the S-Bahn as a unified system. As such, it was for thousands the means of flight to the west and, according to popular history, infested by spies and doubtful characters of all kinds.

4.2.2 Urban transportation in West Berlin

When West Berlin was walled off in 1961 to make it an island in GDR territory, the East Berlin-based Deutsche Reichsbahn (DR) continued to run the S-Bahn on both sides of the city, but as two distinct systems. Lines radiating from West Berlin into the surrounding countryside were blocked at the border with the GDR, and the S-Bahn ring was cut into two independent half-circles by the Berlin Wall (fig. 4.4). The West Berlin section of the system was still allowed to use the north–south tunnel beneath East Berlin, but the only permitted point of interconnection was at Friedrichstrasse Station. Here passengers wishing to pass from the West Berlin S-Bahn or U-Bahn systems to the East Berlin S-Bahn had first to pass through GDR border control.

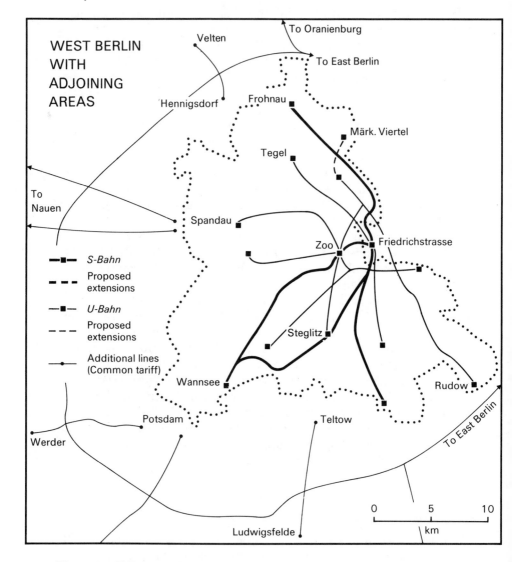

Figure 4.4 S-Bahn and U-Bahn

Friedrichstrasse became the principal crossing point for those entering
or leaving East Berlin.

In West Berlin the building of the Berlin Wall in 1961 was followed
by a popular boycott of the East Berlin-controlled S-Bahn: 'not a penny
for Ulbicht's barbed wire' was the slogan. West Berlin proceeded to
create an urban transport system that totally ignored the S-Bahn; the
bus routes that had supplanted the abandoned tram system frequently

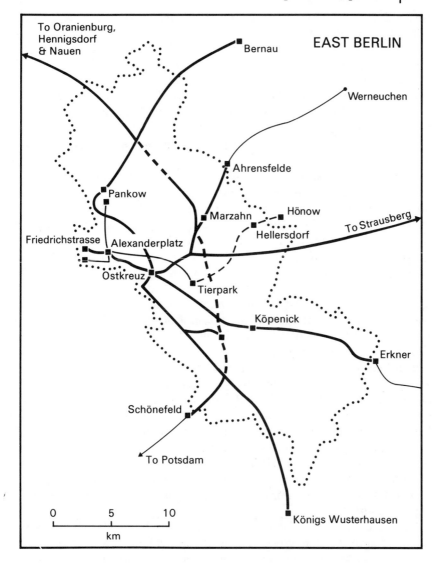

ran parallel to the S-Bahn lines, and were preferred by the population. West Berlin had however the advantage of inheriting the larger part of the pre-1939 U-Bahn system. Two lines actually ran beneath East Berlin and continued to be used by West Berlin traffic, the trains running through darkened ghost stations, where only armed guards occupied the platforms. One west–east section of elevated line across the sector boundary was abandoned, two of its stations becoming respectively the 'Flea Market' and the 'Turkish Bazaar' (see section 8.3), but in general the system was regularly expanded. The emphasis has been on the development of new links to tie the outlying southern and northern portions of West Berlin into the new West Berlin centre. There have been criticisms of the slowness of the system to reach out to serve the great peripheral housing projects of the 1960s and 1970s. The U-Bahn has reached Gropiusstadt to the south, but has got no nearer to Falkenhagener Feld in the west than central Spandau, and will reach Märkisches Viertel only late in the present decade (see section 7.7).

By the mid-1970s the West Berlin S-Bahn was just about on its last wheels. The few passengers had to penetrate into disintegrating stations that looked like sets for horror films of the 1920s, then to ride on the slatted wooden seats of decaying trains through railway land transformed into groves of self-sown birch trees. The East Berlin DR was experiencing financial losses on its West Berlin operations that far outweighed any advantage that control of the system gave it to meddle in West Berlin affairs. From 1975 onwards it attempted to unload the operation onto the West Berlin senate, at first without success. This was partly because of the routine attempt of the GDR government to negotiate directly with West Berlin as if it were an independent free city, partly because it was trying to lease or sell a system that West Berlin held that the East Berlin DR did not own, but over which the western Allies had given operating rights only. Agreement came only at the end of 1983, when West Berlin agreed to take over the system, not from the GDR as a state but from the DR as administrator. It followed that there was no question of 'buying' the system, but substantial payments were promised in respect of services or alleged services provided by the DR, such as running a shuttle link between the Friedrichstrasse entry point and the nearest station in West Berlin (Lehrter Bahnhof) or maintaining the tunnel beneath East Berlin in usable condition.

Possession of an S-Bahn system freed from political stigma will have some advantages for West Berlin. There will be less atmospheric pollution as parallel bus routes are suppressed, and some areas with sparse transport facilities may be better served. But since West Berlin has spent thirty years developing a U-Bahn system without regard to the S-Bahn, there will inevitably be a certain amount of redundancy in the joint system. Even with the help of the federal government, the expenses of restoring the S-Bahn and operating a combined system are such that the city has had to move cautiously. On take-over, a system that had already been reduced by half by the East Berlin DR was further reduced to two disconnected fragments, one running from the interchange at Lehrte Station to the West Berlin centre, the other running from Anhalt Station to the southern borders of the city. The next branch scheduled for reopening also ran to the south. The failure immediately to reactivate the line running through the tunnel under East Berlin to serve the Märkisches Viertel and suburbs in northern Berlin was strongly criticized; passengers burned candles as a sign of mourning in the last train on this line on the night of 8–9 January 1983.

4.3 The Stadtautobahn system

In West Berlin, the plans for public transport were combined with a major programme for the construction of a system of urban motorways. The notion of an urban inner-ring motorway dates back half a century to the plans of Albert Speer; from the same period date the plans for an outer motorway ring, since completed. Clearly, radials linking the two systems were also required, one of which, in the form of the Avus, was substantially available.

The immediate post-war plans for the whole of Greater Berlin involved the attempt to produce a linear central area or *Bandstadt*, linking the traditional East Berlin centre with the emerging West Berlin centre in the Zoo Viertel (see section 7.2). This was to be defined on all sides by a rectangle of 'tangential' roads that would divert traffic from the central area and lead it out to the Stadtautobahn ring. One of these, the *Nordtangente*, was in its course westwards from the Tiergarten largely based on the triumphal axis Strasse des 17 Juni–Heerstrasse that had been developed and extended in the National Socialist period, and so was essentially already in existence. Parts of the *Südtangente* south of the West Berlin centre were completed, largely based on existing streets: it was planned to extend it eastwards through the closely built-up areas of Neukölln and Kreuzberg, south of *Bezirk* Mitte and the line of the present Berlin Wall. So far as possible the Stadtautobahn ring followed the line of the S-Bahn ring, so as to minimize further disruption of urban communities, although this involved time-consuming negotiation with the Deutsche Reichsbahn and considerable additional expense of construction.

After the division of the Berlin administration in 1949, followed by the physical division of the city by the building of the Berlin Wall in 1961, West Berlin was at first reluctant to modify its plans for a tangent/ring system, being unwilling to accord any degree of permanence to the new situation. It was, however, left with plans for half a ring system, a kind of horseshoe with open ends pointing towards non-existent crossings into a totally unresponsive East Berlin (fig. 4.5). With regard to road plans, as to other aspects of long-term planning, the 1971 Berlin Agreement, accepting that Berlin would remain

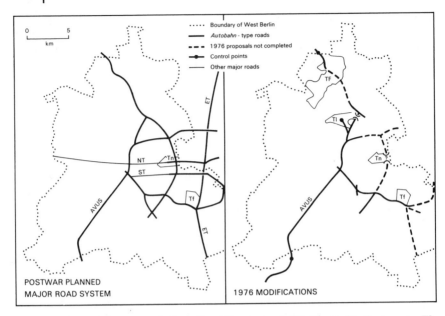

Figure 4.5 The Stadtautobahn. Tn: Tiergarten; Tf: Tempelhof airport; Tl: Tegel airport; NT: Northern Tangent; ST: Southern Tangent; ET: East Tangent; TF: Berliner Forst Tegel. (Source: *Der Tagesspiegel* (Berlin) 16 January 1976, with later modifications)

divided into the foreseeable future, eventually brought a change of plans. In 1976 it was proposed that the Stadtautobahn ring would be closed on West Berlin territory. From it three long-distance links should diverge: the Avus (Route 15) south-westwards to the Dreilinden Control Point and thus to the GDR autobahn system in the directions Hanover, Frankfurt, and Nuremberg/Munich; Route 11 north-westwards across the Tegel Forest to link with the GDR autobahn to Hamburg, via a new control point to be established at Heiligensee; and a transformation of the southern portion of the former Osttangente into a south-east-radial (Route 13) in the general direction of Saxony. Some short feeder sections to the Stadtautobahn were also proposed, while in deference to growing opposition its width was in most sections restricted to three lanes in each direction.

Nearly all of the major road proposals ran into difficulties, mainly through the increasing opposition of environmentalist groups. The building of the Berlin Wall, which deprived the northern and south-eastern districts of West Berlin of their former links across the East Berlin Centre, led to the construction of a 'provisional' north–south link by normal roads across the Tiergarten just west of the Brandenburg Gate. According to the new plans, this was to be replaced or supplemented by a motorway-standard link in the Stadtautobahn ring, possibly in a cutting or even a tunnel. Environmentalist opposition has seen to it that this link has hitherto not been constructed, and it may be that the present unsatisfactory surface link will remain indefinitely (nothing lasts like the provisional). The proposed south-east radial has been a total non-starter, as it depends on GDR co-operation in arranging a new control point, which has not been forthcoming; it has also been opposed by local residents. On the other hand, the GDR has been willing to provide a new control point and link to the Hamburg autobahn in the north-west, but environmentalist groups delayed for years the construction of the necessary autobahn link on West Berlin territory, through the Tegel Forest.

4.4 Urban transportation in East Berlin

By its decision to isolate the 'island' of West Berlin, the GDR had to accept that by so doing it was creating a highly inconvenient traffic obstacle within its own territory. So far as the S-Bahn was concerned, the GDR was left with a number of beheaded commuter lines, which had formerly radiated out north, west, and south from West Berlin. The recently completed, inadequate, and overloaded outer ring railway had to collect up the traffic from these beheaded fragments. As already noted, the rail service to the large town of Potsdam, in the 'shadow' of West Berlin, presented particular problems (see section 4.1.2).

The S-Bahn system is designed to carry the greater part of commuter traffic derived from both within and outside East Berlin. Three-quarters of East Berlin commuters travel for at least the final stage of their journey by S-Bahn. Commuter trains, and even trains used by commuters on the main lines, are halted at the outer terminals of the S-Bahn lines, which now all lie outside the city boundary; for example, many trains from Potsdam terminate at the S-Bahn terminal at Schönefeld Airport. About 65 per cent of suburban bus lines feed into the S-Bahn terminals; only a handful of bus or tram lines cross the city boundary.

A significant recent extension to the S-Bahn system has been the north-east line to Ahrensfelde and Wartenberg, which serves the great peripheral housing developments at Marzahn and Hohenschönhausen. Perhaps the greatest problem is that all trains originating in the East Berlin centre must follow the same line eastwards as far as the old Ringbahn interchange at Ostkreuz, before they can diverge into a fan of routes from north-west to south-west. Travel conditions will be eased by the creation of a new S-Bahn line running right across the city from north to south to the east of Ostkreuz, using existing rail lines to give direct access from Hohenschönhausen and Marzahn to the industrial areas of south-east Berlin and Schönefeld Airport.

The East Berlin S-Bahn is a very socialist system. A monthly 'go anywhere' ticket costs only about £2.00, and five pence will pay for a single journey (by contrast a single journey in West Berlin costs about 50p, although admittedly this allows unlimited changes under certain conditions). The 'honour' system rules; passengers cancel their own tickets and inspectors are never seen; indeed, there has been serious discussion as to whether the income derived from such derisory fares justifies the expense of collecting them. The low (if increasing) rate of private car ownership is reflected in the provision of special compartments for passengers with prams, bicycles, and heavy loads. East Berlin inherited only a small part of the pre-war U-Bahn system, consisting of short north–south and east–west lines intersecting at Alexanderplatz. The second of these lines was extended eastwards to serve the East Berlin Zoo and is now being further extended to the peripheral housing project at Hellersdorf. This should give it much greater traffic and do something to take the strain off the S-Bahn. The main street transport is by tram, but buses are also used, especially in outer districts.

Car ownership is now sufficient to produce problems of parking and traffic congestion. The redesigned East Berlin centre includes a wide inner-urban bypass road on its eastern side (Mühlendamm–Gruner-strasse); in addition, traffic management measures try to contain east–west traffic to the Leipziger Strasse to the south and Wilhelm-Pieck-Strasse to the north (see fig. 6.1). Motorway-standard links to the outer ring autobahn run from Pankow northwards and Grünau southwards; a further high-quality road link to the autobahn runs northwards between the Lichtenberg industrial area and the Marzahn housing project.

5.1 The location of economic activity in Old Berlin

5.1.1 Specialization of areas before 1945

Reference has already been made to the economic transformation of 'Old Berlin' (the historic inner city plus the Electoral 'new towns' of the Dorotheenstadt and Friedrichstadt, see section 1.3) which began after mid-century but particularly gathered pace from 1871 onwards, with the creation of a specialist urban core. The fuller pattern of functional differentiation within this core will be examined below (section 6.2). Of relevance to the present chapter is that, alongside the specialist political, cultural, and social institutions of the capital of Prussia and of the German Empire area there developed parallel specializations of an economic nature.

The 1859–64 building of that essential institution of capitalism, the Stock Exchange (Börse) lay within the inner city, on the banks of the Spree opposite the Berlin Cathedral, but the heart of Berlin banking was further west. The 'street of the big banks' was the Behrenstrasse, which runs parallel to, and south of, the Unter den Linden, although the banking quarter stretched southwards past the Gendarmenmarkt almost to the Leipziger Strasse (see fig. 6.1). On its eastern fringe, at Hausvogteiplatz, lay the 1857 Reichsbank (National Bank), founded to keep watch on them all. The banking quarter formed a transition to the central business district proper, which took up most of the remainder of the Friedrichstadt.

A second activity closely related to the functions of a great capital is the dissemination of information. Newspaper offices were concentrated in a small area of the Friedrichstadt south of the Leipziger Strasse, in fact lying across the line of the future Berlin Wall. While many of the offices were small, the great publishing houses such as Scherl, Mosse, and Ullstein had establishments on a monumental scale. After about

1920 the southern Friedrichstadt also emerged as a distinctive 'film quarter'. South-east of the press quarter, firms connected with the export trade were concentrated on the Ritterstrasse.

A great concentration of wealth and power such as existed in Berlin until the Second World War required the provision of a range of supporting services, many of which were also to be found grouped together in areas of specialization. The western Dorotheenstadt was affected by the presence of the Friedrichstrasse Station, then as now a major intersection of S-Bahn, U-Bahn and tram lines, but until 1945 having more immediate proximity to the command centres of the German state than it has today. Its neighbourhood was dominated by hotels, restaurants, and travel bureaus. Rather similarly, the Potsdamer Platz, in proximity to the Potsdam Station, was a centre of beer restaurants, dance halls, and other places of popular entertainment.

The main centre of high-quality retailing was the Leipziger Strasse, from the Potsdamer Platz to the Spittelmarkt, where it merged into an area stretching as far as Hausvogteiplatz that was renowned for ready-made clothing. Shops selling cloth and dress materials were also crowded into the area.

It must, of course, be remembered that not all the retailing and service specialisms were located in 'Old Berlin'. To name two very different examples, suppliers of plywood were concentrated near the Silesia Station (later East or Main Station), while galleries selling works of art were in the Bellevuestrasse, south of the Tiergarten just to the west of the Potsdamer Platz (Leyden 1933: 148–59). Then increasingly the *Zooviertel*, around the Kaiser Wilhelm Memorial Church in the 'Old West End', was developing as a second centre of high-level retailing and a major centre of entertainments provision (see sections 1.4 and 6.3.2).

The beginnings of Berlin's modern industry are to be found on the fringes of 'Old Berlin'. For example, the clothing retailing of the Leipziger Strasse, Spittelmarkt, and Hausvogteiplatz was backed by a swarm of clothing workshops spreading southwards through the southern Friedrichstadt and the Luisenstadt. The same area had a heavy concentration of printing plants, serving the needs of the capital.

The origins of Berlin's large-scale industry are also to be found in this fringe (see section 1.2). The royal iron foundry to the north-west, outside the Oranienburg Gate, and the private-enterprise plants of Egells, Borsig, and others that clustered near it, are regarded as the origin of Berlin's heavy-machine building, although the larger plants have long since moved further out and their sites have for the most part been replaced by housing. Another industry establishing itself in this quarter was the manufacture of chemicals and pharmaceuticals, particularly associated with the name of Schering. Breweries were a feature of the fringe, a position that the main East Berlin *Getränke-Kombinat* on the Lenin-Allee (Landsberger Allee) still maintains. Then the numerous small courts in this fringe, for the most part built up before 1871, contained a swarm of small plants producing food products, furniture, light fittings, and above all, a wide range of metal objects and machinery. There were also many service activities, such as carriage and (later) motor-vehicle repairing.

5.1.2 Changes after 1945

The whole of this central part of Berlin lay in the area of maximum wartime devastation. The greater part is in the East Berlin *Bezirk* Mitte; only the part of the Friedrichstadt lying south of the Berlin Wall at Checkpoint Charlie and the southern Luisenstadt in *Bezirk* Kreuzberg are in West Berlin. The loss of central economic functions has been almost total; a small Soviet-style socialist state such as the GDR has other priorities than the pre-war capital of a capitalist state of 70 million people. Potsdamer Platz and Leipziger Platz lie flattened under the installations of the Berlin Wall. The Leipziger Strasse and Spittelmarkt are now primarily residential, the Hausvogteiplatz area turned over to official functions. The retail facilities included in the new East Berlin centre in the former inner city tend to have a representational, 'showcase', or specialist function rather than seeking to serve as centres of mass consumption on the capitalist model (see section 7.8). One of the pre-war bank buildings in the Behrenstrasse serves as the GDR state bank, but a socialist state has not much need of banks, so that the banking quarter as such has disappeared. The former Reichsbank has been replaced by the SED Central Committee building and its surrounding car parks, perhaps a symbolic reversal of priorities. The part of the newspaper quarter that was situated in what is now East Berlin has gone, but some printing survives, hard against the Berlin Wall to the south. The north-western fringe was less devastated than the remainder of the area and retains some of its characteristic small-scale manufacturing in internal courts (see fig. 6.1).

South of the Berlin Wall, in the devastated southern Friedrichstadt, printing is represented by a number of plants close to the Wall, including the *Bundesdruckerei* (Federal German Printing Works). The most prominent building is the 79 m high tower of the Springer newspaper and publishing empire, symbolically transferred from Hamburg in the early 1960s to this spot in Berlin's traditional press quarter, on the dividing line of Europe (see fig. 7.2). The nearby Graphische und Gewerbe-Zentrum Berlin was an officially backed attempt to attract publishing and reproductive activities into the quarter, in which it has not been entirely successful. A quite different usage is the establishment here of the halls of the West Berlin wholesale market for plants and flowers. The flourishing clothing industry that once spread in hundreds of workshops over all of this area on the south of Old Berlin has virtually disappeared. The future development of the southern Friedrichstadt was long undecided, with large areas standing empty or devoted to a variety. of low-intensity uses. It now seems likely that the area will be redeveloped with a mixture of residential and business uses (see section 7.4).

5.2 Industry in the Wilhelmian ring

Wrapping round the core of 'Old Berlin' is the broad 'Wilhelmian ring', the product essentially of the growth years 1860–1914 (see section 6.3). It extends approximately from the Customs Wall outwards to the S-Bahn (see figs 5.1 and 6.2). The development and residential characteristics of the Wilhelmian ring will be examined below (section 6.3). It is sufficient to note at this point that (except in the socially superior western sector in Charlottenburg) it was filled by high-density working-class apartment houses, which were built not only fronting onto a monotonous grid of streets but around a series of courts in the interior of the blocks that the streets defined (see fig. 6.4).

These interior courts contained not only residential accommodation but a considerable amount of Berlin's medium- and small-scale manufacturing industry, as well as transport undertakings, repair facilities of one kind or another, and small builders. Although some of this accommodation was abandoned and turned into small residential apartments during the years of the Great Depression, and more has fallen victim to wartime destruction or to post-war urban renewal schemes, what is frequently overlooked is that these undertakings to a considerable degree remain in the same locations to this day.

A survey of thirty-five blocks in Kreuzberg, carried out in the early 1960s in advance of urban renewal, revealed that on an area of 107 hectares there were, in addition to 37,000 inhabitants, no fewer than 2,714 distinct places of employment, with 19,400 employees occupied in hundreds of different trades. Small firms were attracted by the low rents charged in these interior courts, while 60 per cent of their employees lived close enough to walk to work. These are not negligible economic and personal advantages, even if the close intermixing of residential and manufacturing activities, with the disadvantages of noise, atmospheric pollution, and traffic generation offended against the town-planning principles current in the middle of the twentieth century (Hofmeister 1975a: 338–51). From about 1890 this location of manufacturing had come to be accepted as normal to the extent that purpose-built 'rent-factories' were established in the inner courts. K. J. Müller reports from Wedding on an apartment house which has behind it rental factory accommodation arranged around six internal courts, occupied by thirteen firms (K. J. Müller 1985).

By the second half of the nineteenth century industries were emerging, notably locomotive- and machine-building, that required larger areas of land than could be found either in Old Berlin and its fringe, or in the courts of the apartment houses. A process of industrial decentralization was accordingly initiated. Typical in this respect was the 1850 decision of the Borsig machine-building firm, that had originated in the fringe belt just north-west of the inner city, to seek an additional site beside the Spree in Moabit. By 1878 the firm was employing 5,000 workers in Moabit, which had also developed a cluster of other plants devoted to metal processing, machine building, porcelain manufacture, brewing, baking, and pharmaceuticals. Already before the end of the century a number of these plants, notably Borsig, had found the available space too restrictive and had moved even further out (see section 5.3). In part the sites thus liberated were occupied by successor industries, in part converted for residential development (Hofmeister 1981). Similar development took place in northern Charlottenburg, along the Spree upstream from the inner city, and along the railway approaches to the former Potsdam and Anhalt stations through Schöneberg.

Many more sites both for industry and for public utilities were to be found in the vicinity of the S-Bahn ring and generally on the fringe belt of the Wilhelmian ring. The north-west had been established as a major direction of industrial advance since the early nineteenth century, and industrial development continued on this axis through Wedding. Schering chemicals, established here in 1865, had like Borsig originated on the fringe of the inner city. The firm of Schwartzkopff, later Berliner Maschinenbau, established a plant in 1869 to build steam engines, locomotives and road rollers. The vast AEG electrical engineering plants were erected from 1888; the Osram lamp works followed in 1904 (K. J. Müller 1985). Large areas of the north-west were also taken up with railway and storage installations, to which in 1923 was added the Westhafen. Public utilities adjacent to the S-Bahn ring included the Berlin gas works to the north, its prominent brick-built gas-holders recently blown up to make way for the Ernst-Thälmann-Park memorial and housing project, and the vast cattle market and slaughter-house to the east.

The total amount of manufacturing industry contained within the S-Bahn ring, that is to say in Old Berlin and the surrounding Wilhelmian ring, is easily underestimated. In 1925, 70 per cent of Berlin's industrial plants and 63 per cent of its industrial employment lay within the S-Bahn ring. Twelve years after the end of the Second World War two-thirds of the plants and a half of the industrially employed population were still in this location (Hofmeister 1975a: 380–1).

5.3 Industrial decentralization: the outer zone

The years after about 1890 were marked by a process of migration of large industrial plants to sites in Berlin's outer ring, that is the space between the outer edge of the Wilhelmian ring and the 1920 boundary of Greater Berlin, in part even further to satellite settlements beyond the boundary (fig. 5.1). The firms were in search of large areas of relatively cheap land; the sites chosen were mainly on rail and water routes. The largest of the plants formed isolated industrial areas in their own right, strung out at intervals along the means of communication.

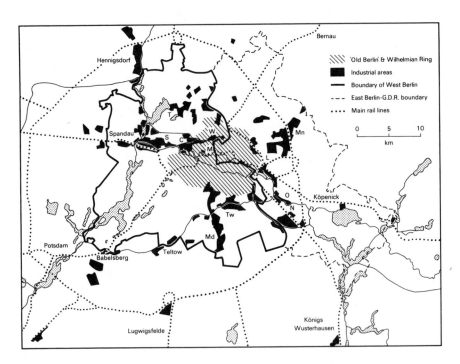

Figure 5.1 Industrial areas of Greater Berlin: M: Moabit; W: Wedding; G: Charlottenburg; S: Siemensstadt; T: Tegel; Tw: Teltow canal; Md: Marienfelde; L: Lichtenberg; Mn: Marzahn; O: Oberschöneweide; N: Niederschöneweide. (Source: Senat von Berlin (West), Senator für Stadtenwicklung und Umweltschutz, Flächennutzungsplan FNP 84, 1984)

They often had their own associated works settlements, but the scale of housing provision was generally too small to remove the necessity for widespread commuting from the older central residential districts (see section 6.4.2).

Yet the move to a site in the outer zone was not the first decentralizing move for a number of these plants. Some writers distinguish four stages in an outwards locational progression from the centre of Berlin. The first sites were in Old Berlin or its fringes, for example to the north-west, outside the Oranienburg Gate. A second, intermediate stage was provided by sites within the Wilhelmian ring, but adjacent to means of communication, such as Moabit. A third stage was provided by sites peripheral to the Wilhelmian ring, along the S-Bahn. A fourth stage was provided by sites in the outer ring.

It has been noted, for example in sections 1.2 and 5.1, that the great locomotive- and machine-building firm of Borsig originated in the early part of the nineteenth century outside the Oranienburg Gate, on the fringe of the inner city. By the 1850s it found itself in need of more room and had established a second producing centre in Moabit, in *Bezirk* Tiergarten (see section 5.2), and by 1878 the site at the Oranienburg Gate had been abandoned. Then in 1898 the Moabit site was considered inadequate and was given up in favour of a vast establishment at Tegel, which was influenced in its location by proximity to the Havel waterway and to the Industriebahn des Kreises Niederbarnim, a railway running from east to west across the north of the city, deliberately created to draw industry from what was then Berlin.

One of the most significant examples of industrial decentralization is provided by the Siemens electrical-manufacturing concern, faced in the 1890s with the need to concentrate the plants that had previously been scattered through Old Berlin and Charlottenburg. The firm selected for this purpose a large site situated astride the boundary between Charlottenburg and Spandau. The site was adjacent to the Spree waterway, but not particularly well served by rail connections. The distinctive red-brick Siemens manufacturing plants and the associated workers' housing settlement have under the title 'Siemenstadt' become a prominent feature of the outer zone of Berlin. Old Spandau, with its armaments plants, formed a further industrial concentration.

A belt of industrial plants is also to be found along the Teltow Canal to the south of Berlin; it includes the extraordinary brick-gothic Ullsteinhaus printing plant of 1926. On the eastern side of Berlin, industrial plants spread along both banks of the Spree, in Oberschöne-weide (*Bezirk* Köpenick) and Niederschöneweide (*Bezirk* Treptow).

5.4 West Berlin after 1945

5.4.1 West Berlin industry after 1945

The post-war history of West Berlin industry is one of disasters from which, however, recovery was often remarkable. It has already been noted that West Berlin's industry, which had already lost about 23 per cent of its pre-war capacity through wartime destruction, lost a further 53 per cent (or 70 per cent of what remained) through Soviet dismantling in 1945 (see section 2.1). During the blockade West Berlin industry almost ceased to function, and emerged economically crippled, cut off from former markets and suppliers in its immediate hinterland, the territory that was in process of becoming the GDR. These had to be replaced by using lengthy transport links to the Federal Republic, which were additionally subject to delays at checkpoints and unpredictable interruptions, thus putting West Berlin industry at a cost disadvantage (see section 2.3).

A further blow came in 1961, with the building of the Berlin Wall. The most obvious effect was the loss of about 53,000 *Grenzgänger*, people who had lived in the GDR and East Berlin but worked in West Berlin. Certain industries were disproportionately affected, notably clothing manufacture, which had drawn heavily on women workers from East Berlin. It was however an encouraging sign for the future of West Berlin industry that an overall improvement in labour productivity rapidly compensated for this loss of workers and for the loss to the labour force of the refugees from the GDR who had previously remained in West Berlin rather than moving on to the 'mainland' Federal Republic. From the mid-1960s the labour force was also supplemented by the recruitment of foreign workers (see section 8.3).

Naturally, West Berlin industry also had to face the problems experienced by industry everywhere, such as technological change and periods of recession in the mid-1970s, following the world oil crisis, and again in the early 1980s. Manufacturing industry (excluding energy production and building) shed 134,400 jobs in the period 1970–83, a loss of 42 per cent, much more severe than the federal average of about 15 per cent. Berlin's traditional leading sector, the manufacture of electrical equipment, was particularly hard hit, losing 48,000 jobs, or 48 per cent of its 1970 total, although remaining (with 52,800 workers in 1984) the most important branch of Berlin's industry. The other traditional pillar of Berlin industry, machine building, was similarly affected (49 per cent), while the third of the main branches, food and drink, lost 27 per cent of its jobs. The once famous Berlin clothing industry had the greatest proportionate losses of all, and in 1984 was operating with only 22 per cent of its 1970 workforce (Kluczka 1985). Although other economic sectors, such as services or public administration, were not affected by such losses (see section 5.4.4), industry's performance was reflected in the overall West Berlin employment loss of 98,900 jobs in the period 1970–84 (10.5 per cent). Unemployment rose from 5,000 in 1970 to 83,600 in 1983, with particularly rapid increases in the mid-1970s and again in the early 1980s, when numbers nearly doubled (*Berliner Statistik*, October 1985).

Isolation and insecurity have been a factor behind a number of industrial losses. Some old-established Berlin firms have either moved their entire operation to the Federal Republic or have established substantial productive capacity there (e.g. Siemens). Any production of even marginal military significance is forbidden by the occupying powers. Admittedly, there has in recent years been some reverse movement of plants from the Federal Republic (e.g. BMW motorcycles from Bavaria, Nixdorf computers from Paderborn), attracted by financial inducements and excellent West Berlin facilities for technical education and research to set up in Berlin, but all too many new implantations are merely branch plants, simple production lines. Policy-making, research, and development are located elsewhere. Such plants can disappear almost overnight, given a change in economic climate. Particularly damaging has been the tendency of firms, even where retaining some productive capacity in Berlin, to move their head offices to the Federal Republic (e.g. Siemens, AEG-Telefunken). By voting with their feet in this way, firms have indicated both a lack of confidence in the long-term security of West Berlin and a consciousness that the national centres of decision-making have moved elsewhere. Very few of the great Berlin firms have remained true to the city; among those that have are Eternit (building materials) and Schering (chemicals and pharmaceutical products). A rare example of a reverse head-office move is provided by the Springer Press and publishing firm, but here the motive was ideological.

The salvation and reconstruction of West Berlin industry would have been impossible without the energy, ingenuity, and steadfastness of its people, but equally impossible without external assistance, initially from the Marshall Plan but essentially from the federal government. The general objective of economic assistance is that living conditions in West Berlin shall not fall below those of the Federal Republic. In addition to direct federal payments equivalent to the general scheme of income equalization between states (*Länder*) to make up what would otherwise be a crippling deficit in the city's budget, three broad categories of assistance to the economy can be distinguished (see section 2.3). Assistance to transport and communications is particularly important, since it helps to remove the cost penalty, if not the time penalty, inherent in West Berlin's isolation.

Especially since the 1971 Berlin Agreement, the government of the Federal Republic has been able to negotiate a number of deals with the GDR government to improve access, including the buying-out of various GDR exactions on transit traffic and payments for the improvement of autobahn and rail access to the city. More direct development assistance takes two forms under the special Berlin legislation known as the *Berlinförderungsgesetz*. In the first place, the operation of Berlin firms benefits from substantially reduced rates of income tax, corporation tax, turnover tax (*Umsatzsteuer*), and business tax (*Gewerbesteuer*). Second, new investment in Berlin benefits from tax-free subsidies of up to 25 per cent, according to the nature of the project. Where investment is exclusively for research and development, the subsidies can reach 40 per cent. Berlin firms also benefit from favourable rates of write-off on new investment.

Vital though this assistance has been, it has been subjected to some criticism. The various aids to investment were not restricted to industry, and it has been claimed that too much was devoted to prestige office buildings, high-priced apartment blocks that made no contribution to solving the city's housing problem, or luxury hotels (although the latter at least have a role to play in the city's attempts to attract tourist guests and conferences). Such projects have even been accompanied by a degree of scandal (see references to the Steglitzer Kreisel, section 7.5). Even in the field of industry, support for new investment seems to have attracted a number of capital-intensive industries making little impact on the employment situation; for example, the manufacture of cigarettes produces a high output when measured in money, and doubtless gratifyingly high taxes for the city authorities, but generates relatively few jobs.

Another way in which the federal government assists Berlin is by steering orders from ministries and other government organizations to the city. For example, the pre-war presence in Berlin of the central offices of the postal system and of the Deutsche Reichsbahn meant that the city received many orders for industrial equipment from these organizations (indeed, the requirements of the railways lay at the very beginning of the Berlin machine-building industry). The federal government now tries to see that Berlin firms continue to receive orders from these and similar organizations, although their successors are now based elsewhere. The assistance given to West Berlin by the siting in the city of various federal administrative offices will be dealt with below (section 5.4.4).

The main financial assistance to new investment in Berlin necessarily comes from the federal government, but the West Berlin government is also extremely active in the field. Financially, its role is mainly one of supplementation, for example in the form of low-interest guaranteed loans for persons starting up in business for the first time. Such persons can also benefit from a range of advisory services. The city also has a role to play in the admittedly secondary but still important matter of the provision of industrial accommodation. Small- and medium-sized firms can be offered rented accommodation in what are called *Gewerbehöfen*, usually reused sites of industrial firms that have gone out of production. Larger areas of land can also be made available; in 1968–72, for example, the city developed as an industrial estate the land of the former municipally owned farm at Marienfelde, on the southern fringe of the city.

In 1981 the SPD regime that had ruled Berlin for so long collapsed in a welter of debt, financial scandal, house occupations, and street riots. In the new CDU-dominated senate under Richard von Weizsäcker (subsequently to be federal president) the senator for economic affairs was Elmar Pieroth, a businessman of original ideas. Emphasis was switched from indiscriminate support to any economic enterprise, whatever its long-term benefit to the Berlin economy, to the encouragement of innovation and high technology. Of particular importance was the creation of the Berliner Innovations- und Gründerzentrum (BIG) on the former AEG land in Wedding, combining the experience of the Technical University with leading industrial firms to help new projects in the high-technology field. An adjoining area of industrial land has been reused to make a 'science park' specifically aimed at new small and medium-sized enterprises. Financial assistance is now concentrated on support for research and for the industrial introduction of new processes. Locational emphasis is on the recycling of abandoned industrial sites to accommodate the new industries, so as not to draw further on West Berlin's scarce land resources.

By 1984 virtually all industrial branches were showing an upward trend in the value of output, although continuing improvements in labour productivity meant that the trend was not immediately reflected in employment figures. Innovation and progress has been particularly marked in electronics and in fields such as automation, laser technology, and fibre-optics. Optimists have even spoken of a second *Gründerzeit*, the equivalent of the explosive period of innovation that Berlin experienced in the period from 1871 to 1914 (Kluczka 1985). From late 1984 industrial employment at last began a hesitant rise.

The 1971 Berlin Agreement, following which it was possible to do something to alleviate the problems of access, also greatly improved the situation with regard to security. Firms can now perceive West Berlin as a location where they may expect to remain undisturbed. What can no longer be believed is that firms located in West Berlin can in any foreseeable future expect to benefit from a wider market provided by the reunification of the city, or indeed of Germany.

Allied with increased security is a changed perception of Berlin as a place in which to live. Investment by the city in good schools, well-equipped universities, housing, public transport, leisure facilities, or museums can be as significant economically as industrial subsidies. Certainly, there seems to be an increasing willingness of highly qualified people to come from the Federal Republic to work in Berlin, and of Berliners with similar qualifications to stay in the city.

5.4.2 Energy supply

West Berlin, in its isolated situation, suffers from the high cost of energy, giving its industry higher costs than in the 'mainland' Federal Republic. The public electrical supply in Berlin is in the hands of the BEWAG, the Berliner Kraft- und Licht AG (the organization has as its head office one of the surviving monuments of Weimar architecture, Fahrenkamp's 1931 Shellhaus, on the Landwehr Canal south of the Tiergarten). Mention has already been made of the way reconstruction of the dismantled 1928–30 Kraftwerk West (later Ernst-Reuter-Kraftwerk) went on even during the blockade (see section 2.7), the necessary machinery being carried by air. West Berlin successfully made itself independent of electric power supplies from the GDR, and thus was able to survive the cutting off of the city in 1952 from the GDR distribution system. Other power stations (in part combined with district-heating plants) are at Charlottenburg (opened 1900), Moabit, Spandau-Oberhavel, Steglitz, Wilmersdorf, Lichterfelde, and Rudow.

The operation of West Berlin as an 'island' system has economic disadvantages. Reserve capacity has to be very much larger than in a normal city with interconnection to a regional or national distribution network. The West Berlin plants have also to function on coal (80 per cent) and oil (20 per cent), expensively transported to the city; under normal conditions it would almost certainly be more economic to produce at coalfield or waterfront locations outside the city. The need to insert plants into the fabric of a crowded city also means that they must be restricted in size, which is an operational disadvantage, somewhat offset by the environmental gain that they can also provide district heating.

The necessity to build power stations within the boundaries of crowded West Berlin produces considerable environmental problems, both visual and with regard to air pollution. The increasing demand for energy led in the mid-1970s to proposals for an additional 600 MW coal-fired plant as far away as possible in the Spandau Forest on the GDR border, but this would have involved the destruction of thousands of trees, and there was vehement opposition from environmental groups.

A clutch of alternatives to the proposal were considered at the time. The idea of an atomic power station was rejected, not so much on environmental grounds but because a minimum of two plants would have been necessary to assure continuity of supply, and this would have vastly exceeded demand. Connection to the Federal Republic's network across GDR territory would apparently not have been opposed by the GDR, provided that enough hard currency changed hands. A Soviet proposal was that the West Germans should build an atomic station at Kaliningrad (Königsberg) in the present territory of the Soviet Union, which would be paid for by electricity delivered on a power line built across Poland and the GDR to the Federal Republic by way of Berlin. The proposal appears not to have found much favour in the GDR, which countered with an offer to sell power to West Berlin from its own network, possibly associated with the building by the West Germans of a nuclear plant on GDR territory, near Magdeburg, which would again have been paid for by deliveries of electricity. Nothing came of any of these schemes, all of which must have incited considerable doubts in West Berlin regarding the likely security of supply. In the meantime West Berlin has again a connection to the GDR network, but the quantity of power transferred is not large. In the end it was decided that additional modern capacity would be built adjacent to the existing Ernst-Reuter plant west of Spandau, following the current West Berlin principle of trying not to take in new sites for industrial purposes when existing ones can be used. Even so, the plant was delayed for a number of years by opposition in the courts.

Power-station and industrial coal and coke comes mainly from the Federal Republic, but the GDR supplies the brown-coal briquettes that are still burnt in the domestic stoves of older properties. About three-quarters of West Berlin's petrol, diesel, and heating oil comes from the GDR Schwedt refinery, mainly by rail. This dependence is not considered to be excessive, as there is sufficient spare capacity in the transport links with the Federal Republic to replace any shortfall in supplies.

As with electricity generation, Berlin had by the 1950s made itself independent in the manufacture and supply of town gas. Under the control of the Berliner Gaswerke (GASAG), West Berlin gas supply was concentrated in only two works, at Mariendorf and Charlottenburg. The old coal-based methods were replaced by gas making on the basis of oil. By 1985, however, under conditions of détente, West Berlin secured a spur-connection from the pipeline supplying Soviet natural gas to the Federal Republic by way of Czechoslovakia. Eventually, a year's supply of gas will be stored in sandstone rocks beneath the Grunewald, not only to ensure security of supply, but to allow for the very great variation in demand for gas for domestic heating purposes between winter and summer. It is hoped that natural gas will replace brown coal and oil in domestic heating, thus reducing environmental pollution. West Berlin would have liked to have the gas transmission line (and any eventual long-distance electricity line) led through the city, so that an interruption to the city's supply would have meant an interruption to the supply of every consumer further west, but understandably the GDR would have none of this. Because of the high capital costs involved in building a pipeline across the GDR and changing the domestic distribution system, a dramatic fall in prices is not expected in the short run, but in the long term the new gas supplies should have at least a steadying effect.

5.4.3 The structure of West Berlin industry

As already noted, the early development of Berlin manufacturing was stimulated by the demands of the army for clothing, leather goods, and weapons, of the administration for paper and printed matter, and of the court and aristocracy for luxury goods such as silk, carpets, porcelain, and objects in gold and silver (see section 1.2). To some extent the state intervened directly in manufacturing, not only with regard to military needs, such as arms manufacture (e.g. in Spandau and Potsdam) and the creation of an iron foundry outside the Oranienburg Gate in the early years of the nineteenth century, but even in meeting the needs of the court for consumer goods (*Kgl. Porzellanmanufaktur*, 1763). Even with the departure of the court, many of these commodities continued to be in demand in a great capital city which combined the presence of an element in the population able to afford luxury articles with an increasing mass market.

The evolution of modern industry in Berlin followed along lines not dissimilar from those of other great European capitals, such as Paris, and which reflect fairly simple principles of industrial location. The city stands in an area virtually devoid of raw materials other than water and a few building materials such as sand, gravel, and cement. Even before 1945 manufacturing industry had accordingly to be based on the transformation of raw materials or semi-finished products that are easily transported from elsewhere. In so far as production was not for the Berlin market itself, the resultant products had to be of a nature and value that could bear the cost of transport back to markets outside the city. Berlin industry characteristically drew upon the skills of its workers and its excellent facilities for technical education and research to concentrate on the later stages of an ever-extending chain of production. There was always a tendency for branches of production initially developed by the skills of Berlin entrepreneurs and workmen to be lost to cheap-labour areas elsewhere. This happened in the nineteenth century with textile production, after 1961 with the West Berlin clothing industry, more recently with some types of electrical and radio equipment. At least until the Second World War, the ingenuity of Berlin's entrepreneurs and workmen was always able to keep the city ahead.

Three branches have dominated the evolution of modern machine-based industry in Berlin. Machine textile production was the first to develop, taking the market opportunity offered by the English blockade during the Napoleonic Wars. When markets were opened again to English goods, the industry migrated to lower-wage centres of production outside Berlin. Textiles were then replaced as an industrial leader by machine building, which held this position until 1871, when the new industry of electrical engineering took over.

Today industrial production in West Berlin is dominated by four major branches. Measured in terms of employment, the manufacture of electrical and electronic equipment comes first, followed by food and drink production, machine building, and the chemical industry. Measured in terms of value added by production, the food and drink industry is well in the lead, followed by electrical and electronic equipment, chemicals, and machine building (Tables 5.1 and 5.2).

The manufacturing of electrical and electronic equipment employs 35 per cent of West Berlin industrial workers and accounts for 11 per cent of output by value. The relatively high labour intensity of this

Table 5.1 *West Berlin: gross inland product and gross value added, 1985*

	DM millions	%	1970 = 100
Agriculture and forestry	107	0.2	105
Productive industry, total	16,163	33.7	101
Energy production	963	2.0	129
Manufacturing	12,247	25.5	108
Building and construction	2,953	6.2	76
Trade and transport	6,861	14.3	101
Service enterprises	13,343	27.8	164
Total enterprises	36,474	76.0	118
State, private households, and			
non-profit-making organizations, total	11,494	24.0	147
State	9,815	20.5	142
Private households, etc.	1,679	3.5	183
Total value added	47,968	100.0	123
Total gross inland product	50,216	–	121

Source: Berliner Statistik, August 1986.

Table 5.2 *West Berlin: employment and output in major industrial branches, 1981–5*

	1981	1982	1983	1984	1985
Employment					
Metal construction	4,045	4,193	4,071	3,892	4,234
Machine building	18,331	17,811	16,239	15,129	14,937
Electrical equipment, electronics	61,380	57,519	53,199	52,838	56,301
Misc. metal products	4,270	3,750	3,446	3,427	3,860
Chemicals	10,955	11,039	11,121	11,493	11,855
Printing and reprographic	6,388	5,882	5,888	5,516	5,363
Clothing	4,939	4,224	3,963	3,660	3,553
Food and drink	22,545	21,794	20,992	20,839	20,519
Total manufacturing	176,560	167,800	159,616	157,539	162,572
Output (DM millions)					
Metal construction	494	514	653	655	681
Machine building	2,026	2,075	1,924	1,869	2,149
Electrical equipment, electronics	3,871	4,021	4,903	6,059	4,978
Misc. metal products	494	511	499	555	662
Chemicals	2,246	2,367	2,578	2,766	3,000
Printing and reprographic	647	651	653	698	722
Clothing	798	814	878	893	890
Food and drink	14,748	15,960	17,878	20,166	21,744
Total manufacturing	32,651	34,682	38,993	43,714	46,301

Source: Statistisches Jahrbuch Berlin 1985.

branch, especially given the advanced technical qualifications required of many of the workers, makes this a particularly valuable activity for the city.

The development of the manufacture of electrical equipment in Berlin is historically related to two firms. The artillery officer Werner Siemens developed a satisfactory indicator-telegraph and was charged in 1847 with the building of a telegraph line between Berlin and Potsdam. In the same year he joined with the technician Johann Georg Halske in the foundation of Telegraphenbauanstalt Siemens & Halske, which from initial workshops near the Anhalt station moved to factories in Charlottenburg and then to the Siemens works that gave a name to a whole Berlin district on the borders of Charlottenburg and Spandau. The Siemens firm was responsible for a whole series of innovations, including the introduction of electric tramways in Berlin, automatic telephone exchanges (1908), developments in radio amplification (1926), and teleprinter systems (1928).

Wartime destruction, post-war Soviet dismantling, and the uncertainty of the Berlin situation led to the transfer of the head office to Munich and a proportion of productive capacity to Munich, Erlangen, and elsewhere in the Federal Republic, but the firm with its 26,000 employees remains the largest Berlin firm in this branch. The Berlin establishment in Siemensstadt, especially the famous 'Dynamo-Werk', with its concentration on generators and electric motors for steel works and ships, went through some difficult times in the early 1980s, and the firm was accused of turning its back on the city. The mid-1980s, however, saw it again investing in new capacity in Berlin, notably in fields such as automation and optical-fibre transmission. Reasons for the change of attitude appear to have included the post-1971 political stability of the city, the high standards of Berlin technical education and research, and the new attractiveness of the city as a place of residence for highly qualified younger people. The new developments involved the complete demolition of the 'Werner-Werk', once one of the jewels in the Siemens crown. Part of the Siemens land is being redeveloped as an 'industrial park'.

The other major firm, the Allgemeine Elektricitäts-Gesellschaft (AEG) was founded later, in 1883, after Emil Rathenau had obtained licences to use the discoveries of the American Edison. The firm concentrated on the field of electricity generation and electric tramways. Subsequent interest in the field of radio was marked by the 1941 merger with the firm of Telefunken; since the Second World War AEG-Telefunken has had its head office in Frankfurt on Main, but has maintained important productive installations in West Berlin. The firm has various plants scattered throughout the city, in Wedding, Moabit, and Reinickendorf, so makes less of a visual impact than Siemensstadt.

Until 1945 AEG productive capacity had been spread throughout central and eastern Germany, and all but 10 per cent was lost. For example, the very large 'outer ring' plant at Oberschöneweide (*Bezirk* Köpenick) is now in East Berlin. The interest on capital raised to replace this lost production was a drain on the resources of the firm. Then, particularly in the early 1980s, the firm was hit by the reduced demand for electrical equipment, being almost driven into bankruptcy. It gave up the use of some of its buildings in crowded Wedding, which were difficult to adapt for modern production; they are in part used by the Technical University for the support of technical innovation (see section 5.4.1) and by other industries (e.g. Nixdorf). The Osram light-bulb firm and other units were sold to Siemens. Modern new buildings were opened in 1985 in Marienfelde, on the southern outskirts of the city, for the application of electronics in industrial production, and at Spandau for railway equipment. By the mid-1980s AEG was said to be again 'in the black'; Berlin, with 7,150 employees, was still the largest centre for AEG activities, measured by employment.

Siemens and AEG have a variety of subsidiaries in Berlin. Among other producers, apart from Telefunken plants (linked with AEG), electric lamp production is important (Osram in Wedding, with in 1985 a vast new plant in Spandau). Another major firm is Standard-Elektrik-Lorenz, owned by the American ITT, with works on the Teltow Canal; like so many other firms, it experienced market problems with its short-range radio equipment in the early 1980s, and was subsequently re-equipped to make terminal screens and related products. A number of other firms, among them IBM and Triumph-Adler have contributed to the very significant recent expansion of electronics and related fields such as data-processing and office machinery. The Paderborn firm of Nixdorf in 1985 built a new plant on the former AEG land in Wedding.

In spite of its historical importance, the manufacture of machinery and vehicles now employs only 9 per cent of all industrial workers and accounts for only 5 per cent of output by value. As noted above, it was particularly the demands of the railways that led to the establishment of a group of firms producing locomotives and other equipment on the north-west fringe of the inner city, outside the Oranienburg Gate. In the later nineteenth century they moved out, often finding larger sites in the north-west sector through the Wilhelmian ring, in Wedding, Moabit, or northern Charlottenburg. Then either after making such an intermediate halt, or directly, firms seeking really large sites moved from the 1890s into the outer zone or even beyond (see section 5.3). One favoured area was Marienfelde, on the railway running southwards through *Bezirk* Tempelhof. In the process of time firms have greatly adapted their productive range; Schwartzkopff, for example, originally a major locomotive builder, was in the 1970s producing coin-operated vending machines at its plant in Wedding. Firms that failed to adapt to changing demand have often ceased to exist.

Even the great Borsig firm has had severe problems. The successive moves that led to the vast Tegel site have already been described (section 5.3); a further steel- and rolling-mill plant yet further out in the satellite settlement of Hennigsdorf is now in the GDR. Locomotive building was given up in 1918 and the firm turned to the manufacture of steam turbines, castings, and pressings. A sequence of changes of financial control indicated the firm's difficulties in the inter-war years. The Second World War brought great damage through bombing and Soviet dismantling, and production was only restarted in 1950. Electric steel production was based on locally available scrap; products included boilers, turbines, compressors, valves, equipment for the chemical industry and refrigerating plant. At the beginning of the

1970s the firm with over 3,300 employees was still one of the largest in West Berlin, but economic difficulties continued.

The production of rolling stock both for the normal rail system and for the U-Bahn can be regarded as a continuation of a traditional Berlin branch of production. Berlin has at various times attempted to begin the production of motor vehicles, but without much success. There is some production of excavators and other specialized vehicles and of diesel engines. An interesting development was the 1969 decision of the Bavarian BMW firm to concentrate all its motorcycle production in Spandau; a second highly automated plant was added in 1978. The manufacture of lifts (elevators) and of refrigeration equipment are other Berlin specialities.

The food-and-drink industry was essentially related in its development to the mass market provided by the 4.3 million inhabitants of Berlin (1939), supplemented by an 'upper class' of persons able to pay high prices for luxury items. Recovery after 1945 was slow, partly because Marshall Aid and similar assistance tended not to be generously available for consumer-goods industries, partly because for a number of years food and drink items were offered by the GDR at dumping prices. In recent years the processing of food and drink has shown a dramatic rise in importance, making it by far the most important industrial branch in Berlin in terms of the value of output, and second in terms of employment. The relationship to the local West Berlin market is still strong; whereas West Berlin as a whole sells about 85 per cent of its output outside the city, about half of the output of the food and drink industries is internally consumed.

It is of course in line with the basic propositions of industrial location that a productive activity such as brewing, which relies primarily on a ubiquitous raw material (water), should be market-orientated. Berlin possesses famous names (e.g. Schultheiss) and at the Technical University has one of the Federal Republic's two schools of brewing. Obviously, the manufacture of soft drinks of all kinds is similarly market-orientated. Berlin with its high-income consumers has long been a centre of distilling, although since the end of the 1960s its tax preferences have been eroded and the industry has come increasingly under competitive pressure from producers at home and abroad. Berlin chocolate manufacturers have important markets outside the city; this is even more true of cigarette manufacture, which since the 1950s has become heavily concentrated in Berlin and delivers the overwhelming proportion of its output to the Federal Republic outside the city. West Berlin is the largest centre in the Federal Republic for coffee roasting.

Among other industrial branches, the chemical industry is third in terms of the value of output, fourth in terms of labour employed. It follows the general Berlin pattern in concentrating not on the mass production of basic chemicals but on highly processed pharmaceutical products and cosmetics. Like other major Berlin firms Schering originated in the middle of the nineteenth century in the fringe belt north-west of the inner city, in the *Bezirk* Wedding, still the location of its main plant. It is one of Berlin's major employers and has other plants in the Federal Republic and abroad. At the end of the 1970s it erected 'Schering-Stadt', a complex of administrative, social, and research buildings, providing a gleaming contrast to the generally run-down environment of Wedding.

The textile industry has a significant place in the early history of factory production in Berlin, but plants migrated away in search of cheaper labour in Saxony and Silesia. The industry is no longer of importance in West Berlin, although there are a few extremely modern post-war plants, especially in relatively capital-intensive branches such as knitwear. The days are long gone when, before the Second World War, Berlin was responsible for 80 per cent of German production of women's outerwear. This industry was effectively destroyed by a series of blows. After the destruction of the war years, the division of the city left much of the industry in East Berlin. Because elaborate equipment was not required by the then characteristic small-scale establishments, recovery after the blockade was at first rapid. Then in 1961 came the building of the Berlin Wall, which overnight deprived the West Berlin industry of 8,000 mainly female workers, and strengthened the trend to move production to the Federal Republic or even to distant low-wage countries. Another factor in the decline of this branch is West Berlin's loss of capital functions; it is no longer the predominant centre of wealth that it once was. In terms of high fashion, cities such as Düsseldorf, Frankfurt, Stuttgart, or Munich are at least of equal significance. To the extent that the industry survives, it is no longer as a swarm of middlemen and outworkers spreading from Spittelmarkt southwards through the Luisenstadt into Kreuzberg. There has been a concentration into larger plants producing for the chain stores of the high streets rather than for individuals, and these are dispersed throughout West Berlin.

The relationship of printing and newspaper publishing to capital functions has already been mentioned (see section 5.1). The loss of such functions, together with wartime destruction and the division of the city, were bitter blows to these branches. Many forms of printing require close contact between press and customer, which put the West Berlin industry at a disadvantage when seeking work in the Federal Republic. Whereas in 1939 Greater Berlin had 35 per cent of printing employment in Germany, and the area of the present West Berlin 26 per cent, the share of West Berlin had by 1970 sunk below 5 per cent (Hofmeister 1975a). As part of federal government help to Berlin the

Bundesdruckerei (federal, formerly state, printing works) has been maintained in Berlin. The city's role as a capital and as a centre of academic excellence meant that it was also an important centre of book and map production, although because of Germany's political history it never had the national dominance of Paris or London. The post-war dispersal of refugee publishing firms from the traditional book centre of Leipzig benefited cities of the Federal Republic rather than West Berlin, which by the 1970s was responsible for only 7 per cent of new titles (Hofmeister 1975a).

West Berlin naturally has a number of other manufacturing concerns with more than local reputation, among which may be mentioned the *Königliche Porzellanmanufaktur* (KPM), established in 1763. This obviously falls into the category of court-related porcelain works, of which more famous examples are Meissen and Nymphenburg. Its products are marked by the sceptre from the electoral arms of Brandenburg printed in blue beneath the glaze. The works were almost totally destroyed in the Second World War, but restored in 1953–4 with the most modern equipment.

5.4.4 The tertiary sector in West Berlin

In value-added terms, the tertiary sector overtook the secondary sector in West Berlin from the beginning of the 1970s; by 1984 its share of the West Berlin total was 58 per cent. In employment terms, the change had arrived some years earlier; by 1984 nearly 70 per cent of West Berlin employment was in the tertiary sector.

West Berlin's retailing activities, together with related entertainments activities, are dealt with below in connection with the West Berlin centre (section 7.2) and secondary centres (section 7.5). A retailing firm, Hertie, is actually West Berlin's second largest private employer, following Siemens. As well as its 'own-name' stores it owns the Wertheim stores and the Kaufhaus des Westens (KaDeWe), claimed as Germany's biggest department store. Clearly, Berlin is no longer the financial and banking centre that it was as Reich capital before 1945 (see section 1.3). In 1945 what remained of the banking centre on the Behrenstrasse was closed by the Soviet occupying power. Only official or co-operative banking institutions were allowed. In West Berlin the commission established to see through the first currency reform in 1948 developed into the Berlin Central bank, which as the central banking institution of one of the constituent states of the Federal Republic was from 1957 incorporated into the Deutsche Bundesbank (the Federal German Central Bank). The West Berlin segment of one of the few initially permitted banking organizations, the Berliner Stadtkontor, became in 1950 the Berliner Bank, the largest of the

Berlin public-sector banks. It was joined by Berlin subsidiaries of the big three banks of the Federal Republic (Dresdener Bank, Commerzbank, Deutsche Bank) and numerous purely Berlin institutions. Their present function relates essentially to Berlin business; the role of banking and financial capital city has been taken over by Frankfurt. Banking and finance accounts for only 4 per cent of Berlin's gross product and 2.5 per cent of its employment.

Of much greater importance in employment terms is the public service, which in 1982 overtook manufacturing industry to be responsible for about a quarter of West Berlin employment (see table 5.1). The figure is high by comparison with other major West German cities such as Hamburg, and the West Berlin government has been accused of padding the pay-roll to reduce unemployment, and perhaps to reward political friends. It must be remembered, however, that the federal government has deliberately tried to compensate Berlin for the loss of civil-service employment that it once had as a capital city by establishing in the city some of its less vital and non-military agencies. More than 44,000 persons are employed in about fifty federal institutions (1986). The largest, the Insurance Office for Federal Employees, has over 12,000 workers. Other notable agencies are the Federal Agency for the Control and Supervision of Cartels, the Federal Public Health Agency, and the Federal Printing Works. This increased federal presence is regarded with distrust by the Soviet Union and the GDR, who claim that the federal government has no place in West Berlin. Protests were particularly vehement over the installation in West Berlin of the Federal Environment Agency (Umweltbundesamt), which was held to be effectively a federal ministry.

Another way in which the West Berlin government endeavours to increase the city's income and employment is through the encouragement of the city's tourist and conference trade. West Berlin has a somewhat exceptional position as a tourist centre, in that the city alone is the target. Visits to the surrounding environment are limited to one-day trips to East Berlin, to see a very different political system in action, or to Potsdam and the park of Sanssouci. Nevertheless, Berlin's urban tourism has increased steadily in recent years; twenty-five years ago it ranked seventh among the cities of the Federal Republic for the number of visitors, now it ranks second. The duration of visits tends, however, to be short; the average tourist feels that he has 'done' West Berlin in two to three days (Vetter 1984).

Over 1.7 million people visited the city in 1985, and bed-nights totalled 4.5 million in 1985; both figures should be increased by an estimated one-third to allow for those staying with friends or relations. Although foreign visitors are prominent in museums and entertainment facilities, they make up only about 18 per cent of the total and account for only 22 per cent of bed-nights; clearly, the Federal Republic is the main source of tourists. The tourist trade has benefited from the growth of short breaks in addition to the main holidays. Organized groups come by plane or bus to occupy the hotels at weekends or holiday periods, when the normal business or official use is low. There is also an appreciable occupancy of the lower grades of hotels and pensions by students, trainees, and even by skilled industrial and building workers, who live in the city during the week and return to the Federal Republic at weekends.

Hotel development has benefited from the same federal incentives to investment as has industry; indeed, there was at one time talk of overprovision of luxury accommodation, but now underprovision is the more usual allegation. There has been a marked trend for the number of hotels and guest houses to diminish (445 in 1967, 339 in 1945), with the loss of the small and medium-sized units, but for the number of beds available to increase (8,000 beds in 1960, 22,300 in 1985). At times of really major events visitors have had to be accommodated in East Berlin or even in Hanover, commuting by plane! The West Berlin government helps with information services and tries to encourage the provision of low- and medium-cost accommodation suitable for families, young people, and workers.

Of particular importance for the city is the attraction of international congresses, exhibitions, and sporting events, where the possible financial rewards are great, but competition is not just nation-wide but world-wide. The Exhibition Halls beneath the radio tower in western Charlottenburg have been improved and extended. The collapse of the Congress Hall in the Tiergarten was obviously a blow, but it was rebuilt for the centenary year 1987. Fortunately in 1979 the Berlin Austellungs-Messe Kongress GmbH (AMK) opened its new Internationales Congress Centrum (ICC). The siting of the building reflects changing planning policies in West Berlin; instead of being placed with the Congress Hall and other post-war public buildings close to the sector boundary with East Berlin (see section 7.4), it is well to the west side of the West Berlin centre, adjacent to the Exhibition Halls. Here it rises like a stranded ocean liner, its conference halls and all imaginable related facilities laced together by moving staircases. It enables Berlin to claim to be the leading centre for international congresses in the Federal Republic, ranking sixth in the world following Paris, London, Brussels, Geneva, and Vienna.

5.5 Economic balance-sheet

By the second half of the 1980s, Berlin's economic performance could in many ways be regarded as satisfactory. The overall gross product of the city was climbing steadily out of the trough of 1981–2, the rate of recovery being faster than that of the Federal Republic as a whole. This applied to virtually all sectors except, intermittently, building and construction, always a volatile element. Although manufacturing industry was declining in relative importance, it still accounted for just over a third of the city's creation of wealth and was taking on workers again. Progress was particularly marked in the high-technology field. The city had a favourable balance of visible trade both with regard to the Federal Republic and internationally. Only with regard to the GDR did West Berlin buy more than it sold, but the totals were smaller (see section 2.6).

A particularly high rate of expansion was consistently maintained throughout the 1970s and 1980s by the service sector, responsible for about a quarter of the gross product. Some other fields, such as financial or public services, expanded less rapidly, while making substantial contributions to the economy. All this has meant a steadily expanding number of persons in employment. Unfortunately, due to increasing numbers of persons entering the labour market, new immigrants, women, and young people, there has been little impact on total unemployment. There is also the familiar situation of a shortage of skilled persons in key branches co-existing with a substantial pool of persons insufficiently qualified to take up the opportunities offered. Attempts are made to bridge the gap not only by training schemes but by tempting skilled persons to migrate from the Federal Republic, with some success (see section 8.2).

One long-term anxiety of West Berlin's industrial leaders concerns the city's present policy of preventing the further use of 'green field' sites, in the interests of preserving the environment. It is feared that the recycling of abandoned industrial sites will not be sufficient to meet industry's future needs.

5.6 The East Berlin economy

5.6.1 The structure of East Berlin industry

East Berlin, like Berlin as a whole in the pre-war Reich, continues to be the largest single industrial city of the GDR. Measured in terms of employment, however, there has until very recently been a tendency for the industrial importance of the capital to decline slightly, owing to the planned development of industry in the remainder of the Republic. Between 1960 and 1983 Berlin's share of GDR industrial employment fell from 6.7 per cent to 5.3 per cent, in absolute terms a decline of about 7,000 workers. Since 1983 there has been a turn round. Between 1983 and 1985 East Berlin's share of GDR industrial employment rose again to 5.3 per cent; in absolute terms there was an increase of 14,600 jobs in the period from 1980 to 1985. Production is destined to continue to rise at above the average GDR rate at least until the end of the 1990s, underpinned by a rate of industrial investment that in the 1980s was double that of the previous decade (Kehrer 1986).

The reason for this 'back to Berlin' trend lies in a changed direction of GDR industrial policy. In the 1970s the GDR government decided that the country could not advance simultaneously on all fronts of advanced industrial research. It left certain fields, such as telecommunications, for other states in the Soviet bloc to advance, deciding that the GDR would concentrate on the high-technology fields of biotechnology and microelectronics. The latter in particular meant Berlin. GDR research activity is heavily concentrated on Berlin, which measured by the numbers employed has 20 per cent of the research capacity of the GDR. The city has one scientific worker for every 4.3 industrial workers, as against a GDR proportion of 1:16. This predominance is unlikely to diminish; between 1986 and 1990 half of the planned investment of the Academy of Sciences (under which is grouped a constellation of research institutes) will go to Berlin (Kehrer 1986). It is expected in the GDR that scientific research, indeed learning of all kinds, should make a strong contribution to the needs of the state. In Berlin it is required that local combinations of research and of productive capacity shall be developed for the purpose of industrial innovation, especially in the key high-technology areas.

The nature of Berlin industry provides an appropriate basis for such developments (tables 5.4 and 5.5). As in West Berlin, there is a traditional concentration on the manufacture of electrical equipment. This has provided the basis for developments in electronics, particularly in the advanced high-technology fields such as microelectronics. For example, East Berlin does not engage in the mass production of chips but specializes in the production of chips and other components that are tailor-made for specific applications, used not only by electronic industries in Berlin but in other parts of the GDR. Berlin specializes also in optical-electronics and the making of optical-fibre cables. Electrical equipment and electronics, the city's leading industrial branch, account for 31 per cent (1984) of the city's gross industrial production (16 per cent of GDR production). The electronics industry also combines with East Berlin's second largest branch of manufacturing, machine building (15.8 per cent of East Berlin manufacturing, 3.8 per cent of the GDR) to produce computerized machine tools and industrial robots. Chemicals rank third, the branch concentrating on pharmaceuticals and fine chemicals, but in gross output terms the industry is overshadowed by the producers of basic chemicals in the east and south of the GDR. The emphasis on labour-intensive and high-technology

Table 5.3 *East Berlin: employment by economic sector, 1984*

	Employment (000s)	% of total employment	Women as % of total
Manufacturing	165.9	25	39
Productive crafts			
(excluding building)	16.5	2	38
Total	182.4	28	38
Building	54.5	8	17
Agriculture and forestry	6.9	1	51
Transport and communications	73.3	11	35
Trade	97.4	15	66
Other productive branches	41.5	6	55
Non-producing activities	204.6	31	64
Total employment	660.5	100	50

Source: Statistisches Jahrbuch Berlin 1984.

Table 5.4 *East Berlin: gross industrial production, 1984*

	Production as a % of:	
	Total Berlin production	Total GDR production
Energy production	10.1	4.9
Chemicals	12.4	4.3
Water	1.2	11.2
Machinery and vehicles	15.8	3.8
Electrical equipment,		
electronics	30.6	15.2
Light industry	15.3	8.5
Food and drink	11.0	4.2
Other	3.6	2.5
Total industry	100.0	5.4

Source: Statistisches Jahrbuch Berlin 1985.

industries is shown by the way in which electrical equipment/electronics, machine building, and chemicals now make up 59 per cent of East Berlin's industrial output, as compared with 43 per cent in 1953.

Berlin's traditional light industry, producing mainly consumer goods, nearly equals machine building in the value of output, accounting for 15 per cent of the industry of the city and 8.5 per cent of the total industrial output of the GDR for this branch. East Berlin is still responsible for a fifth of the GDR's clothing output; furniture manufacture, printing, and publishing are also important. As in all

great cities, food and drink industries provide a significant proportion
of total industrial output in the city without being of great significance
in relation to the GDR as a whole; clearly, the greater part of output is
consumed within the city itself. East Berlin's share in the GDR output
of particular products is given in Table 5.5.

Table 5.5 *East Berlin percentage of GDR production:
selected items*

	%
Colour TV tubes	100
Electric lamps up to 200 W	100
Insulin	100
X-ray film	100
Radio recorders	100
Optical–electrical components	99
Brakes for trains and trams	92
Electrical equipment for industrial plants	53
High-tension switchgear	46
Electrical cables and wires	45
Semiconductors and chips	11

Source: Magistrat von Berlin 1986.

Unlike West Berlin, East Berlin is integrated into the GDR
transmission system, allowing it to be supplied with the bulk of its
electric power from plants on the brown-coal fields to the south. A
characteristic of East Berlin is the reliance for the provision of heat
and hot water on a series of large district-heating plants, which also
generate some electricity. Originally they burned Soviet oil, but
supply and foreign-currency difficulties led to their conversion to burn
untreated brown coal, with unfavourable consequences for the level of
atmospheric pollution. This trend is general; in 1985 Berlin's consumption
of imported deep-mined coal was 65 per cent of that in 1980, and of oil
it was 20 per cent of the 1980 figure. Over the same period
consumption of untreated brown coal rose by 406 per cent, of sieved
brown coal by 321 per cent (*Zur Entwicklung Berlins 1970–1985*).

5.6.2 The location of industry in East Berlin

East Berlin in 1945 inherited the same industrial pattern as Berlin as
a whole (see sections 5.1 to 5.3). Much industry, especially small-scale
plants, lay embedded in the courtyards of the fringe of the inner city,
especially to the north and north-west, and of the Wilhelmian ring.
Then, especially after about 1890, industrial decentralization brought

about the growth of large industrial concentrations in the outer ring and even beyond the boundary of the present city.

The present territory of East Berlin has three major industrial sub-regions, each containing about a quarter or a third of industrial employment (see fig. 5.1). It comes as something of a surprise that in spite of considerable outward movement of manufacturing from Old Berlin and the Wilhelmian ring these inner areas contain the largest of the three concentrations (Leupolt 1986). The largest peripheral concentration of industry inherited from pre-war times lies on either side of the Spree upstream from the city. On the north bank Oberschöneweide (*Bezirk* Köpenick) includes the former AEG electrical-equipment plant. Across the river similarly industrialized Nieder-shöneweide (*Bezirk* Treptow) continues south-eastwards through Johannisthal and Adlershof. This south-east sub-region has the disadvantage of being badly placed with regard to access from the major housing projects that are currently under development east and north-east of central Berlin (see section 7.7). Completion of the proposed cross-city S-Bahn link should ease this situation (see section 4.4).

Poor access is not characteristic of the main area of post-war industrial development in Lichtenberg and Marzahn. This eastern sub-region now stands second in terms of employment, benefiting from a theoretically ideal relationship to residential provision. To the west of the new S-Bahn line to Ahrensfelde lie the industrial plants, some of them relocated from sites in the Wilhelmian ring. To some extent there is an outward gradient of skills, from, for example, the Marzahn machine-tool plant to plants handling building materials far out to the north-east. East of the railway is the vast Marzahn housing project. As always with this type of planning, it cannot be guaranteed that those offered dwellings in the housing project also work in the corresponding industrial area, but at least adequate S-Bahn links are at hand, and they will soon be better.

From now on it is unlikely that the pattern of industrial distribution in East Berlin will change very much. As in other cities, it is now accepted that industrial plants situated within the existing urban area should remain, so long as they are not environmentally damaging. The process of technological change is also regarded very differently in the socialist GDR than in capitalist countries. Instead of plants using out-moded technology being quite simply closed, and new plants established elsewhere, the socialist GDR will try to find an alternative line of production to occupy the existing labour force. Another development tending to emphasize existing locations is the attempt to minimize transport costs by building up local productive relationships by agreement between neighbouring firms.

5.6.3 Industry in the agglomeration field

The process of industrial decentralization that began at the end of the nineteenth century did not stop at the borders of Berlin; industrial establishment requiring large amounts of space were established further out on the radiating rail lines. The resulting distribution of industry was very uneven; only certain directions, such as Hennigsdorf–Oranienburg, Erkner–Rüdersdorf or Wildau–Königs Wusterhausen were favoured. Under the socialist principles of proportional economic development the GDR has since 1947 attempted to produce a more widespread distribution of industry, especially along the radials to Bernau and Strausberg that were previously relegated to a commuting population only (Leupolt 1983).

In contrast with the approximately circular agglomeration field of pre-war days (fig. 5.2), the obstacle provided by West Berlin means that the agglomeration field is restricted to the half circle that is readily accessible from East Berlin. The larger industrial concentrations are restricted to the 'inner agglomeration field' immediately bordering East Berlin, where there are four important concentrations accounting for about two-thirds of the field's total employment (fig. 5.3).

One of the clearest examples of the extension of space-consuming industries into the agglomeration field is provided by the northern axis Hennigsdorf–Oranienburg. At Hennigsdorf the former AEG plant concentrates on the building of electric locomotives, particularly for mining purposes, but also for S-Bahn and U-Bahn trains, with a world-wide export. The steel works and rolling mills work primarily on scrap derived from the GDR. Oranienburg produces chemicals; nearby Velten, clay products and refractories.

To the east, Rüdersdorf has old-established industries producing cement and building products; nearby Erkner, chemical industries. To the south, Königs Wusterhausen–Wildau have metal-processing industries and varied light industries. The industrial concentration centred on the medieval town of Bernau is mainly the result of post-war planning, including food processing and the manufacture of components for the Hennigsdorf electrical-equipment industries. More isolated centres in the outer agglomeration field are provided by Ludwigsfelde (motor-vehicle production), Zossen, and Fürstenwalde.

The agglomeration field is linked with East Berlin in a variety of ways. First of all, it provides Berlin with labour; at the end of the 1970s 32,000 commuters moved daily to work in East Berlin's industrial and building sectors. East Berlin also looks to the agglomeration field for purposes requiring large amounts of land, for example the production of building materials, warehousing and storage, sewage farms, and electrical transformer stations (Leupolt 1983). There is also an

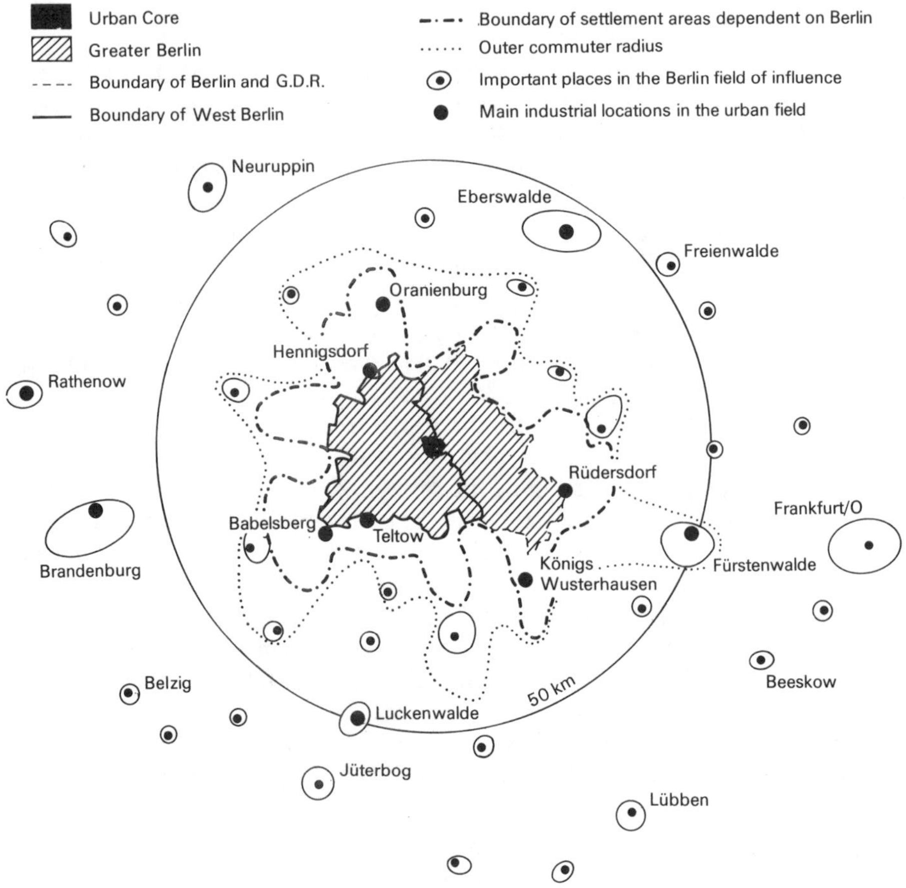

Urban Core
Greater Berlin
- - - - Boundary of Berlin and G.D.R.
——— Boundary of West Berlin

—·—· Boundary of settlement areas dependent on Berlin
······ Outer commuter radius
Important places in the Berlin field of influence
Main industrial locations in the urban field

Figure 5.2 The Berlin agglomeration field, 1937 (Source: Zimm 1961a)

increasing industrial integration. The field already has over a third of all industrial employment in the agglomeration as a whole; it typically delivers raw materials and semi-finished products for further processing in the capital, but the flows are not entirely in the one direction. In a wider sense the agglomeration field provides many food products to Berlin and is much visited for informal recreational purposes. In return, East Berlin provides important educational, cultural, and organized recreational services to its surrounding region.

5.6.4 The tertiary sector in East Berlin

In contrast to West Berlin, East Berlin is a functioning capital. This means that unlike West Berlin, East Berlin offers employment in

ministries appropriate to a state of its size. More than that, it must be
noted that the command positions of the state are occupied by the
Socialist Unity Party (SED), and that at each subordinate level in the
administration, in industry, or other organizations, the SED has a
parallel organization. A further parallel organization is provided by
the trade unions. In addition to the presence in East Berlin of the
People's Army and the frontier police, any part of the GDR, and not
least Berlin, must be expected to have a substantial contingent of the
State Security Police. Because East Berlin is a capital city, it contains
the command institutions not only of these bodies but of a range of
fields in the economy, the arts, and the sciences, for example the
Academy of Sciences on Platz der Akademie (formerly Gendarmenmarkt). Given the highly bureaucratic and militarized nature of the

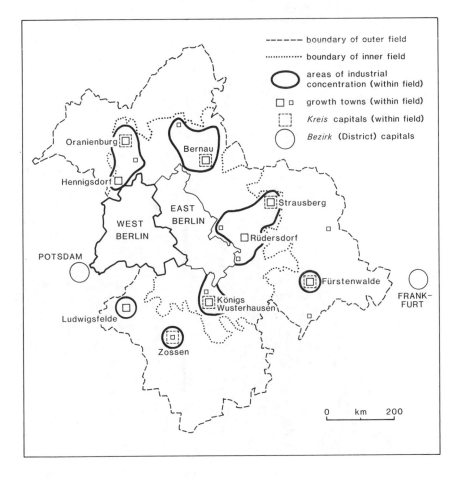

Figure 5.3 The East Berlin agglomeration field (Source: Rumpf *et al.* 1983)

Plate 5.1 GDR tourism is mainly organized on the basis of tourist groups or delegations. Here, one such group visits the Ernst Thälmann monument, north of the East Berlin centre, set in a landscaped new housing project with 1.337 dwellings.

GDR, the employment structure of the capital must be expected to contain a high proportion of civil, police, and military officials. Official statistics give the 'non-productive sector' as accounting for 31 per cent of East Berlin employment, but it is not clear how many of the categories mentioned above are included in this figure.

East Berlin obviously derives income from tourism, but because tourism in the GDR is organized on a national rather than a *Bezirk* basis, statistical information is difficult to obtain. The number of people coming from West Berlin on a daily basis is also not recorded. That tourism in East Berlin has increased in the last fifteen years cannot be doubted. Foreign-tourist visits to East Berlin, whether from socialist countries or from the west, characteristically take the form of 'inclusive tours'; individually arranged visits are much rarer. The typical East Berlin visit normally lasts for only two days, after which the tour is completed in other parts of the GDR.

Many visitors are content to wander round the reconstructed East Berlin centre, to see the contrast between socialist planning and the affects of unrestrained capitalism in West Berlin. The cluster of museums on the 'Museum Island' is a major attraction; the collections rank in world importance only after the British Museum and the Louvre (Bräuniger 1986). A distinctive form of tourism consists of day visits by coach from West Berlin. Being peered at through the windows of glossy coaches gives the present writer, when resident in Berlin, the feeling of being in a human zoo. The day visits often include a restaurant meal or a visit to one of East Berlin's theatres, ranging from the Opera to the new Friedrichstadtpalast for variety entertainment. Many of the best-known theatres could be booked out for two years ahead on foreign-tourist demand alone (Bräuniger 1986).

East Berlin has about 6,000 hotel beds. Some of the more recently erected hotels are directed towards customers paying in western currencies. When West Berlin has important events, such as the *Grüne Woche* (Agricultural Fair) or the *Funkausstellung* (Telecommunications Fair), many international delegations find it easier to book hotel accommodation in East Berlin than in West Berlin (Vetter 1984).

6 | Urban development

6.1 Urban structure before 1945

The Greater Berlin of the 1930s, as described by Leyden (1933) and Louis (1936), had a structure that accorded with all three of the most familiar urban models. It followed the Burgess model to the extent that a pattern of concentric zones could be discerned (Burgess 1925), based however on the sequence of development outwards from the centre, not on social-area differentiation, as in the initial Chicago example. The ideas of Hoyt also appeared to be confirmed by the existence of sectors of contrasting economic status radiating from the inner city (Hoyt 1939), traversing the concentric growth rings and giving a marked East End/West End effect. Finally, although there was only one predominant urban core, such a large urban area necessarily had subsidiary nuclei (a characteristic enhanced by events since 1945), so that the multiple-nuclei ideas of Harris and Ullman also apply (Harris and Ullman 1945).

6.2 Old Berlin

The core of the unified Greater Berlin of 1920 was provided by the 'Old Berlin' (Alt-Berlin) of 1709, which had been created by the union of the original twin town of Berlin–Cölln with the planned seventeenth- and eighteenth-century electoral extensions (*Kurfürstenstädte*). Alt-Berlin was contained within the 1735 Customs Wall (*Zollmauer*) and was broadly identical with the present *Bezirk* Mitte (fig. 6.1). Beginning with the *Gründerjahren* 1870–1914 it was progressively transformed into the Berlin Urban Core (the central business and administrative district, what the German writers call 'the city').

6.2.1 The inner city

The very heart of Old Berlin was the *Innenstadt*, 'inner city', the

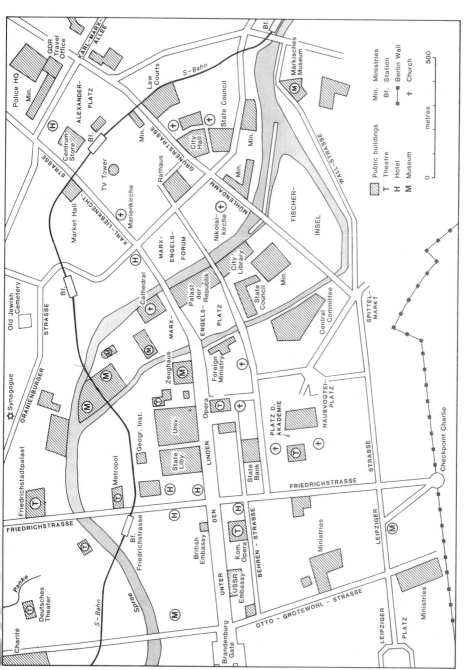

Figure 6.1 The East Berlin Centre

original medieval twin-town of Berlin–Cölln. Until the middle of the nineteenth century it was still set apart from the remainder of Berlin by its ring of seventeenth-century fortifications, and even after their removal it continued to have a distinctive urban morphology and, to a degree, distinct functions. Yet to write about this oldest part of Berlin as it was until the early 1940s is to write about what no longer exists, to a far greater extent than is true of other parts of the present city.

Leyden and Louis have left us a picture of this area as one of transition. It had long been abandoned as a place of residence by members of the official caste, who had moved westwards, and since about 1860 it had also been steadily losing its characteristic lower-middle-class inhabitants through transformation into part of Berlin's Urban Core. Only a few fringing areas were still essentially residential; otherwise the inner city had been invaded by uses such as shops, offices, warehouses, and workshops, as well as by the characteristic small cafés serving the workers. The process was reflected in urban morphology; the two- or three-storey houses built in the seventeenth, eighteenth, and early nineteenth centuries, houses that could be matched in any middle-sized town of the Brandenburg Mark, had been partially replaced by purpose-built commercial buildings. The general effect was described by Louis as unpleasing and disorderly (Louis 1936: 150). Leyden comments that the Berlin *Rathaus* and the more recently built neighbouring *Stadthaus* were lost in a maze of buildings lining an essentially medieval pattern of narrow streets, failing to make much impact on the urban scene (Leyden 1933: 55). From a distance, however, the inner city was still marked out by the steeples of its ancient churches. Functionally, in addition to local administration and commerce, the inner city was part of the Berlin clothing district, which stretched away southwards in a swarm of small workshops.

It is typical of towns and cities to reject to their fringes activities which are either space-demanding or regarded as noxious. Should the town continue to grow, such features may still remain embedded in the urban mass, forming what Conzen called a fringe belt, defining the extent of the town at the earlier stage of development (Conzen 1960). Where a town or city has grown concentrically in a number of stages of development, an equivalent number of fringe belts will be discernible. Such is the case with Berlin (fig. 6.2).

The innermost fringe belt is provided by the emplacement of the seventeenth-century fortifications and related features beyond. The bastions of the fortification were, on dismantling, often replaced by small squares, gardens, or public buildings. For example, the land of the upstream bastion south of the Spree is now occupied by a museum of local history (Märkisches Museum) and a small park containing a

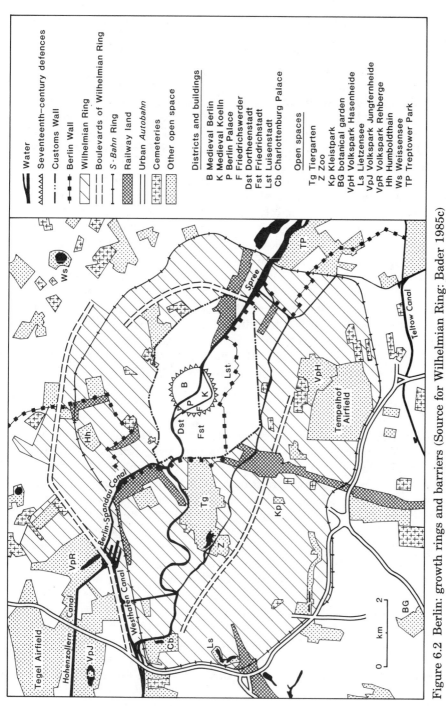

Water
Seventeenth–century defences
Customs Wall
Berlin Wall
Wilhelmian Ring
Boulevards of Wilhelmian Ring
S-Bahn Ring
Railway land
Urban Autobahn
Cemeteries
Other open space

Districts and buildings

B   Medieval Berlin
K   Medieval Koelln
P   Berlin Palace
F   Friedrichswerder
Dst Dortheenstadt
Fst Friedrichstadt
Lst Luisenstadt
Cb Charlottenburg Palace

Open spaces

Tg Tiergarten
Z   Zoo
Kp Kleistpark
BG botanical garden
VpH Volkspark Hasenheide
Ls Lietzensee
VpJ Volkspark Jungfernheide
VpR Volkspark Rehberge
Hh Humboldthain
Ws Weissensee
TP Treptower Park

Figure 6.2 Berlin: growth rings and barriers (Source for Wilhelmian Ring: Bader 1985c)

few remains of the fortifications and bear pits with living representatives of Berlin's symbolic animal.

On the northern side of the inner city, a break in the urban fabric marking the line of the fortifications was in 1882 occupied by the east–west S-Bahn line. The Alexanderplatz marked the point where the Königsstrasse (later Rathausstrasse), then the main south-west–north-east route across the city, passed through the fortifications. A typical fringe feature outside the northern fortifications is the oldest of the Jewish cemeteries, in use from 1672 to 1824, but stripped of its monuments during the period of National Socialist rule.

The fringe belt on the western side coincided with low-lying land on the flood-plain of the Spree, which, being difficult to build on, was of low value and occupied mainly by the gardens of the Berlin Palace. Here human effort has produced a dramatic change in values. The Lustgarten itself, immediately beside the vast Palace, was transformed successively into a military parade ground and a scene for National Socialist demonstrations. Beyond it lay another vast symbol of national authority, the ornate Protestant Cathedral of 1894–1905. Then the whole of what is now the north-western end of the Cölln island was devoted to the building of Prussian National Museums, constructed with great difficulty because of soil conditions. In this way, what was once an unimportant fringe has become perhaps the most important cultural centre of the city (Louis 1936: 150).

6.2.2 The electoral 'new towns'

The position of the palace on the western fringe of the inner city was vital for the development of the emerging metropolis. From the beginning, the open-air distractions and the residential comings and goings of ruler and court were predominantly directed westwards and south-westwards, along an axis parallel to the Grunewald in the direction of Potsdam. In the immediate vicinity of the Palace the way led westwards over the Lustgarten to the hunting preserve of the Tiergarten, later transformed into a park. In the more distant hunting forests along the Havel the elector in 1542 established a hunting lodge 'zum grunen Wald'; today Jagdschloss Grunewald is one of the oldest buildings in West Berlin. The way to it lay in part over swampy ground, so ran on an embankment (Damm); the 'elector's embanked road' (Kurfürstendamm) is now the best-known street in central West Berlin. The banks of the Havel and Spree became peppered with palaces, hunting lodges, summer-houses, and other pleasure buildings. It is true that East Berlin has six palaces of its own and a further one in Oranienburg, but the emphasis is firmly westwards, with four palaces, including Charlottenburg, on the lower course of the Spree, and ten on the Havel, or near to it in the Potsdam neighbourhood (fig. 6.3). To these may be added about another ten great houses of aristocratic families or wealthy capitalists, of which perhaps the best known to non-Berliners is Schloss Tegel, home of the von Humboldt family.

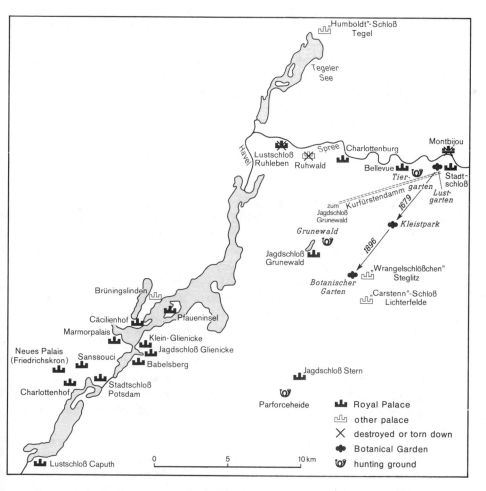

Figure 6.3 Palaces, gardens, and hunting preserves of western Berlin (Source: Hofmeister 1985)

The Botanical Garden has followed a parallel western trajectory. Originating in the Lustgarten at the Palace, it was moved in 1679 to the site of an electoral hop-garden, near Schöneberg on the road to Potsdam; a remnant of this land now forms the small Kleist Park. In

1896 it was moved further south-west on the same axis, to a site of 42 hectares between the road to Potsdam and the village of Dahlem.

This location of the palace on the west side of the inner city, together with electoral perception of the advantages of the west and south-west, and the fact that most of the land in immediate proximity was in the ownership of the elector, led to the initiation of a typical Hoytian sector of preferred residential and other developments. New urban extensions were created that were legally 'new towns', under direct electoral control, and independent of the administration of Berlin–Cölln. The first of these, Friedrichswerder, was founded in 1670 on a relatively narrow strip of land south-west of the island of Cölln, that had been included in the Great Elector's ring of fortifications (fig. 6.2). Directly west of the Berlin Palace, the 1674 Dorotheenstadt was traversed by the Unter den Linden, the type of parade axis characteristic of the princely capitals of the period, leading to the Brandenburg Gate and the Tiergarten. The Unter den Linden was progressively lined by palaces built for various princes and by major public buildings, notably the Opera.

To the south, initially separated from the Dorotheenstadt by a wall, lay the larger Friedrichstadt, initiated by the Elector Frederick III, from 1701 King Frederick I. The new development was characterized by a rectangular grid of wide streets, strongly contrasting with the narrow, crooked streets of the Inner City (see fig. 6.1). The Friedrichstadt had as its major east–west axis the Leipziger Strasse, parallel to the Unter den Linden, and terminating westwards in the Octogon (Leipziger Platz) before the Potsdam Gate. Dorotheenstadt and Friedrichstadt were linked by the Friedrichstrasse, which ran southwards, intersecting the Unter den Linden and the Leipziger Strasse, to terminate in the Rondell (later Belle-Alliance-Platz, later Mehring-Platz). A centre for the Friedrichstadt was provided by the creation of a market-place, the Gendarmenmarkt (subsequently Platz der Akademie), from 1701 graced by twin French Protestant and German churches, and from 1821 by Schinkel's Schauspielhaus (Prussian State Theatre).

Land in the towns that was not already owned by the electors was acquired by exchange or sale, if necessary under compulsion. This complete control enabled the court officials to determine not only the street plan but the height and appearance of the buildings, which had to conform to the general plan. Essentially, these were residential developments, intended for officials of the court and persons of similar standing. There is evidence that the electors did not find it easy to fill up their new developments, and a certain amount of pressure on persons connected with the court, as well as inducements in cash and in kind, were necessary before plots were taken up.

The new towns were originally built up with respectable, two- or three-storey stone houses, except that the western edge of the Friedrichstadt, along the Wilhelmstrasse, was from the end of the eighteenth century selected for the building of palaces for princes and the nobility. From the second half of the nineteenth century everything changed, as the former residences of Berlin high society were absorbed into a markedly asymmetric central area (urban core), which to the east barely reached across the old inner city to the Alexanderplatz, but westwards extended to the edge of the Tiergarten. The original houses were, for the most part, replaced by massive new administrative and commercial buildings, which could visually be more easily absorbed by the wide streets than in the congested inner city. As in all great cities, there were areas of specialization within the central area, which have been described by Leyden and Louis.

The eastern part of the Dorotheenstadt, adjoining the Berlin Museums, developed a specialist function relating to learning, with the university, the state library and, in the streets north of the Unter den Linden, many other academic institutes, such as the Geographical Institute in the Universitätsstrasse.

The western side of the Friedrichstadt underwent a very different development, its palaces being transformed into the centre of national government, for which the street name 'Wilhelmstrasse' (now Otto-Grotewohl-Strasse) became synonymous. Among the official buildings were the palace of the president of the Reich, Bismarck's Reich Chancellery, Hitler's new Chancellery (behind which was built the wartime bunker in which he died), the Foreign Office, Goebbels's Propaganda Ministry, and south of the Leipziger Strasse, the vast hulk of Göring's Air Ministry, which still survives. In addition, specialist retailing and service areas emerged (see section 5.1).

6.2.3 The boundary of Old Berlin

At the middle of the nineteenth century the area hitherto described, together with some irregular suburban development north, east, and south of the inner city, made up the total built-up area of the emerging metropolis of Berlin. The whole was enclosed within the Customs Wall of 1735, which had been designed to leave generous space for urban expansion; this space was still not entirely filled up at this stage. In the west the wall coincided with the boundaries of the Friedrichstadt and Dorotheenstadt, and thus approximately with the edge of the built-up area, except for a scatter of villas along the southern fringe of the Tiergarten. In the north it was reached and, in places crossed, by the belt of suburbs fringing the inner city, but southwards beyond the planned development of the Luisenstadt, and to the south-east, there were stretches of land not built up until after 1860. The Customs Wall is accompanied by the features of a second fringe belt; where the built-up area did not reach the wall, then the fringe features were sometimes located inside it rather than outside. They were carefully examined by Louis (1936: 155–61).

The fringe features are at their clearest in the west, where the Customs Wall and the edge of the built-up area coincided; here the gardens of the palaces along the Wilhelmstrasse provided a wide green corridor where, ironically enough, today runs a corridor containing the complex series of obstacles that we know as the Berlin Wall. Because railway building took place at this stage of the city's growth, and because the customs authorities placed difficulties in the way of rail construction inside the Customs Wall, the great rail termini, often with associated goods yards, constitute one of the most typical features of this second fringe. The adjoining Anhalter Bahnhof and Potsdamer Bahnhof in the west, the Hamburger Bahnhof and its successor the Lehrter Bahnhof in the north-west, the Stettiner (Nord) Bahnhof in the north, and the Görlitzer Bahnhof in the south-east are all just outside the Customs Wall. In the east the existence of unbuilt-up land enabled the Schlesischer (Ost) Bahnhof to be established within the Wall.

This second fringe zone is also marked by a ring of cemeteries, belonging to the various Berlin parishes as well as other groups, such as the Jewish community. Prisons (Moabit) and hospitals (Charité, Bethanien) are also characteristic. Barracks, another use, are exemplified by the splendidly named 'Kaserne des 1. Garde-Dragoner-Regiments, Königin von Grossbritannien und Irlande, of 1850–3 in Kreuzberg, still surviving to remind us of a time when the British and Prussian royal houses were not as sundered as they were to become in 1914. Military training grounds were another feature, the largest at Tempelhof south of Berlin (subsequently Berlin Tempelhof Airport). The origins of Berlin's large-scale industry are also to be found in the fringe (see section 5.1).

6.3 The Wilhelmian ring

The next zone, called the *Wilhelminische Ring* (Wilhelmian ring) because it was created during the reigns of Kaisers William I and William II, essentially corresponded with the population explosion of the *Gründerzeit*, when from 1871 Berlin grew to be a capital city on a world scale (see section 1.3). The major lines of development reflected the city plan drawn up by James Hobrecht in 1862, which was designed to govern the expansion of the city to an eventual population of 4 million people over the space of a hundred years. In detail the type of housing provided was governed by the *Bau-Polizei-Ordnungen* (regulations of the Building Department), which were not restricted to the territory of Berlin city alone, but extended into at least the inner areas of surrounding municipalities. This zone of very high-density residential development, with much intermixed industry, has as its approximate inner boundary the Customs Wall, lying somewhat outside it in the north, somewhat within it in the south and east, where areas were not built up before 1860. The outer boundary extends to the line of the S-Bahn ring railway in the west, south, and east, somewhat beyond it in the north (Bader 1985a). It coincides with the *Bezirke* Friedrichshain, Kreuzberg, Tiergarten, and Wedding, together with the inner portions of Schöneberg, Wilmersdorf, Charlottenburg, Neukölln, Prenzlauer Berg, and Lichtenberg.

Hobrecht's planning had much in common with that of Hausmann, except that it did not involve the extensive demolition of pre-existing property that was considered necessary in the much more extensive fabric of pre-industrial Paris. The ideas of both Hobrecht and Hausmann had common roots in the formal town plans of the seventeenth and eighteenth centuries, as exemplified in Berlin by the Dorotheenstadt and Friedrichstadt. In Berlin, relatively wide radials led out to a ring boulevard, never entirely completed, which to the north defined the Wilhelmian area, and to the south lay within it (see fig. 6.2). The circus (*Sternplatz*), where a number of roads converge, is another characteristic element of the Berlin plan, like that of Paris. In Berlin, such convergences are often marked by a church, visible from afar. The most important ones were of stone, like the Kaiser Wilhelm Memorial Church at the head of the Kurfürstendamm, but in the working-class districts they are mostly in a massive red-brick gothic (irreverently nicknamed 'God's power houses'). The scarcity of public gardens and playgrounds is a frequently criticized feature of the plan.

6.3.1 *The Mietskasernen (rent-barracks)*

The intervening areas were then divided by a monotonous grid of streets, producing a structure of extremely large blocks, between 100 m and 250 m deep, three or four times the size of those in the Friedrichstadt. The streets were constructed so as to carry the heaviest traffic, not just for access purposes, involving relatively high costs, which had to be borne by the developers. It has been suggested that it was to meet these costs that the developers filled their plots to the total extent allowed by the building regulations, although the cynical might suggest that they would have maximized their returns by doing so anyway.

Development essentially took the form of apartment buildings, normally five stories high, but behind the street façade building continued around a series of interior courts, so as to use to full capacity the extremely deep plots into which the excessively large blocks were divided (fig. 6.4). It was not uncommon to find three or even five courts in succession. Two-thirds (in central districts three-quarters) of the plots could be built up, and the size of the interior courts was dictated by the minimum that the fire brigade would accept for the deployment of their rescue apparatus. Until 1868 the minimum areas was 28.5 sq. m, with a minimum distance between the residential buildings of 5.3 m (Schinz 1964). The minimum area was raised to 40 sq. m in 1868, 60 sq. m in 1887 and 80 sq. m in 1897. To cram in the maximum number of families, the working-class apartment blocks were frequently subdivided into extremely small one- or two-room apartments, and in addition it was common for cellars and basements to be inhabited. In 1890 they made up 7.7 per cent of the total housing stock; some were still occupied in the difficult years after the Second World War. A combination of low incomes and the large numbers of people flocking into the city to find work led to the widespread presence of sub-tenants and lodgers, generally men. In working-class areas at least, sanitary standards were low; apart from the problem of the airless and sunless interior courts, shared toilets were provided on landings or in the courts, baths were unknown, while the walls of cellar dwellings frequently ran with damp.

These were the *Mietskasernen* (rent-barracks), the characteristic home of the Berlin working class. Urban densities were in these circumstances extremely high; Leyden recorded that in 1925 numerous census-enumeration districts had residential densities in excess of 1,000 per hectare. The extremely high residential densities were justified on the grounds that only in this way could reasonably priced dwellings be provided for the poor in an area of high land prices (although it could reasonably be argued that it was the very possibility of building to high densities that allowed the high land prices in the first place). The system was also sometimes justified on the ground that the range of rents between the 'good' apartments facing the street and the worst rooms in cellar or garret encouraged social mixing.

As noted, the interior courts contained not only residential blocks but a considerable amount of Berlin's medium- and small-scale manufacturing industry (see section 5.2). An even more curious feature was the siting of schools in the interior of blocks, presumably for the same reason of a search for cheaper land away from the main streets.

By the second half of the nineteenth century industries were emerging, notably locomotive- and machine-building, that required larger areas of land than could be found either within Old Berlin and its fringe or in the courts of the *Mietskasernen*. Firms in these industries increasingly established their plants on the Spree upstream or downstream from the inner city (Moabit, North Charlottenburg) or on the fringe of the Wilhelmian ring, preferably along the Ringbahn (Wedding).

6.3.2 The old West Berlin

The ring of *Mietskasernen*, with a high density of mainly working-class residents and much intermixed manufacturing was, however, interrupted

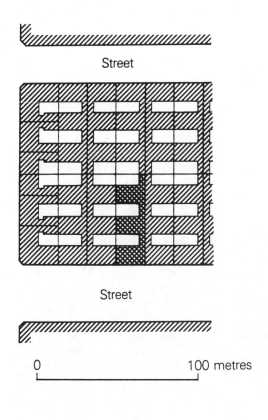
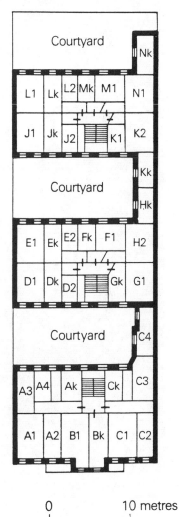

Figure 6.4 *Mietskasernen* in Berlin
(Source: Schinz 1964)

on the western side by a characteristically Hoytian sector of higher-status development. The immediate interruption was provided by the expanses of the Tiergarten, the fringes of which became something of an extension of the Wilhelmstrasse, with government buildings such as the Reichstag, the Ministry of the Interior, and the War Ministry. It had been the intention of the National Socialist government to build a great triumphal axis running south from the Reichstag to a new South Station in Schöneberg, lined with grandiose buildings, but apart from some demolitions that were soon swallowed up in the greater destruction of the war, little was achieved. It was also proposed to concentrate embassy buildings in the villa quarter south of the Tiergarten, where a number were already established; the war-damaged Italian and Japanese Embassy buildings remain as monuments to this plan.

West of the Tiergarten lay Charlottenburg, a municipality which gained in prestige and amenity by the presence of Schloss Charlottenburg and its park. Charlottenburg's fiercely guarded civic independence was not ended until its inclusion in Greater Berlin in 1920. At least in its south-eastern portions along the Kurfürstendamm and the Kantstrasse, it emerged as a sector of apartment buildings devoted to the middle classes and upper middle classes. The superior status of the apartment houses along the Kurfürstendamm was evidenced not only by their elaborate façades but also by the fact that they were the first residential buildings in Berlin to be equipped from the beginning with lifts.

The presence of a relatively high-income population was one factor in the evolution, even before 1945, of a degree of division of labour with regard to central-area functions between the overcrowded Berlin Urban Core ('City') and Charlottenburg (see section 1.4). The district around the Kaiser Wilhelm Memorial Church had good access; as early as 1892 seven tram-routes converged on the Zoo rail and S-Bahn station, and by 1913 there was also a station of the new U-Bahn in close proximity. Charlottenburg was also selected as the site of a number of important public institutions, including the Charlottenburg

Plate 6.1 *Mietskasernen* in Kreuzberg, West Berlin, *c*.1980. Characteristic five-storey apartment houses of the Wilhelmian ring, aligned on a monotonous rectangular street grid. The tall arched doorways lead to internal courts, with further apartments and sometimes industry. The small basement and corner shops and pubs that served the teeming working-class population before 1945 are now mostly closed and barred. The district now has a large Turkish population together with old people and other disadvantaged members of the German population. Housing of this type is being progressively subjected to a renovation programme, in part in relation to the 1987 International Building Exhibition.

Technical High School (now Technical University) and, somewhat further out, the Exhibition Halls and the Haus des Rundfunks (Broadcasting House).

This pre-1945 West Berlin was decreasingly a sub-centre, more and more existed in its own right; the drawing-power of the West was underlined by the inclusion of the term in the titles of many activities (Theater des Westens, Kaufhaus des Westens). Particularly in the inter-war years, the Kurfürstendamm evolved into a centre of high-quality and luxury retailing, whereas the older centre in the Leipziger Strasse offered a wider assortment of goods. Above all, it was a major entertainments centre; in the inter-war years the concentration of a group of first-run cinemas gave it a clear lead in the new and developing field of film. The emergence of this second urban core was to be of great significance in the evolution of Berlin after 1945 (Hofmeister 1980).

6.4 The outer zone

The generous territory accorded to the Greater Berlin of 1920 meant that the dense-packed Wilhelmian ring was in turn surrounded by a wide zone of much lower settlement density, stretching outwards to the new city boundary. This zone contained, and still contains, very large areas of forest and lakes to the west and south-east, as well as considerable areas of agricultural land, of garden colonies, and (especially to the north-east) of land for sewage-irrigation (*Rieselfelder*). The zone included a number of villages, some of which, like Lübars in the north or Marzahn in the east, were still essentially agricultural, as well as two towns of medieval foundation, Köpenick to the east and Spandau to the west. Its development was to contrast with that of the Wilhelmian ring, which had grown by successive accretions to the existing urban mass. The outer zone developed a much more open pattern of distinct cells of settlement or industry, which at least initially were separated from each other by undeveloped land.

6.4.1 Villa settlements

It must not be thought that the 'Burgess' zones of Berlin developed in strict temporal sequence. Even at the inception of Greater Berlin, the outer zone contained much scattered, low-density urban settlement, especially along the routes radiating from the centre. Much of this had evolved contemporaneously with the building of the Wilhelmian ring but was of contrasting character. In particular, this was true of the villa settlements that began to emerge from the 1860s onwards. Typically, they were created by developers on poor agricultural land or through erosion of the forest fringes and consisted of single-family houses standing in gardens. Whereas the planning of the Wilhelmian ring looked towards the Parisian model, the villa colonies were strongly influenced by English single-family houses. While some of the earliest developments had the ideal of providing a relatively economical alternative to high-density urban living, provision for the wealthier groups soon came to be the dominant objective; even the Kaiser had a villa built in the exclusive Grunewald settlement.

The villa settlements were heavily concentrated in a Hoytian-type sector running from the edge of the Wilhelmian ring south-westwards to Potsdam (fig. 6.5). The rail lines to Potsdam formed the principal axis of these settlements, giving commuter access to central Berlin, while the inhabitants benefited from proximity to the recreational areas of the Grunewald and the Havel. Housing for higher-income groups continued to be built in this sector in the inter-war years, and it remains the preferred residential location for post-war West Berlin. There were some other villa settlements elsewhere in the outer ring, notably in the proximity of the forests and lakes to the south-east, but they were fewer and smaller.

Contained within the south-western sector, at Dahlem, was an area of farm land in state ownership, only part of which was released for building villas. Some of the rest was used in 1912–16 to begin building the Dahlem Museum, which was the only museum building in West Berlin to come through the war undamaged, and which now forms one of West Berlin's three museum clusters. Land was also used for various state-supported research institutes, one of which, the 'Kaiser-Wilhelm-Gesellschaft zur Förderung der Wissenschaft', now 'Max-Planck-Gesellschaft', was in 1948 to provide initial accommodation in its institutes for the newly established Free University. Proximity to these institutions further emphasized the residential attractiveness of the sector.

Similar to the villa settlements were a number of 'garden cities', which were not truly independent settlements of the Howard type but rather more or less well-planned garden suburbs using the informal English garden-city style (which had been strongly influenced by the model of picturesque German small towns, as portrayed in the works of Camillo Sitte). The essentially suburban nature of 'Gartenstadt Frohnau', on the far northern fringe of West Berlin, is shown by the way that its most imposing and centrally placed building is the S-Bahn station. It was laid out in a romantic, anti-urban style, with winding streets, numerous green spaces, trees everywhere, and detached, single-family houses set in their own gardens; the post-war period, with its shortage of building land within West Berlin, was greatly to modify this picture by the division of gardens to create plots for

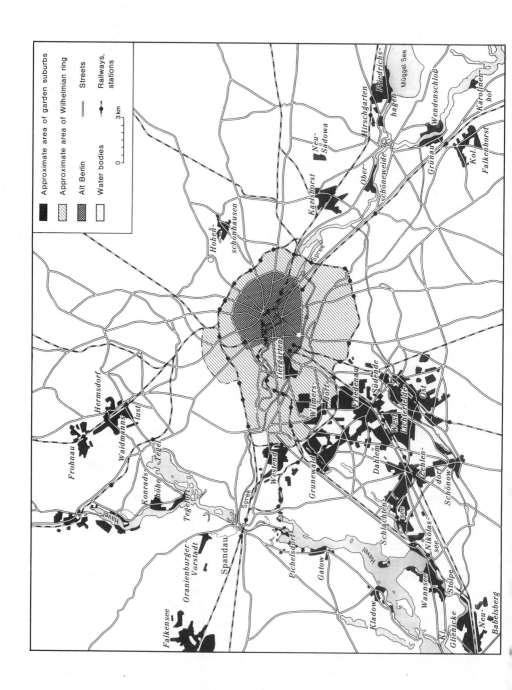

	Approximate area of garden suburbs		Streets
	Approximate area of Wilhelmian ring		Railways, stations
	Alt Berlin		
	Water bodies		

0   3 km

additional houses, and by the building of apartment blocks (K. J. Müller 1985).

6.4.2 Decentralized industry and works settlements

It has already been noted that from the 1890s some of the larger Berlin industrial concerns were attracted by the availability of considerable areas of relatively cheap land to decentralize their plants to new sites in the outer zone (see section 5.3). Clusters of industrial plants are thus another characteristic of the landscape; they were mainly located on the waterways and rail routes radiating from the city (see fig. 5.1). Initially, the workers had to commute from their former places of residence within the S-Bahn ring; at the new Siemens plant on the borders of Charlottenburg and Spandau, transport links were initially so bad that workers were faced with a walk of between 2 km and 4 km from the nearest rail stations or tramway halts (Voll 1985a). Some had to be ferried by boat from an S-Bahn station on the other side of the Spree! In view of their remote locations, some of the firms followed a well-established German tradition by beginning to build housing for their workers, although it should be emphasized that in most cases the greater part of the labour force continued to commute.

The quality of housing provided varied considerably. Borsig at Tegel appears to have been content to provide fairly basic apartment blocks, but elsewhere more interesting developments were to emerge. 'Gartenstadt Staaken', west of Spandau on the boundary with the GDR, was built during the First World War for workers in the Spandau armaments plants. With its small, mostly single-family gabled terrace houses and large private gardens, it was influenced by English garden-suburb style and by Dutch models, but also reflected a wish to return to the village or small-town settlement forms of a pre-industrial Germany. Residential building at 'old' (pre-1945) Siemensstadt began in 1904, but its great expansion came in 1929–32, when it took its place as one of the characteristic Berlin settlements of the Weimar period, to which we now turn.

6.4.3 Housing developments in the inter-war years

The availability of land in the outer zone also provided the opportunity for the extraordinary publicly backed building programme of the Weimar years, which supplemented and greatly extended the settlement-building activities initiated by industrial firms such as AEG, Borsig, and Siemens from the end of the nineteenth century. Early creations of the architects who dominated Berlin's settlement activity in the 1920s at first followed broadly in the garden-suburb spirit, as in Bruno Taut's 1913–14 'Gartenstadt am Falkenberg' at Grünau, in the south-east of Berlin, although even here the bold use of colour on the exteriors of the single-family houses showed the aesthetic influence of Expressionism (Schulz and Gräbner 1981).

By the 1920s the architectural style was essentially that of the 'modern movement', where the external form of buildings was held to be determined by their internal function and by the industrialized methods used to create them. Plain façades and the bold use of colour predominated externally, and there was an almost universal use of flat roofs, which were somehow regarded as 'progressive'. On both economic and aesthetic grounds there was a movement from the garden-suburb cottage style to a more 'urban' layout. Whereas most of the settlements had their quota of single-family terrace housing, there was an increasing tendency to favour long slabs of three- or four-storey apartment buildings. The austerity of this layout was, however, tempered by the liberal provision of gardens and open space, and the occasional use of curved rather than linear blocks. Social provision was not forgotten; the settlements were provided with shops and sometimes with other facilities, such as a cinema, a kindergarten, or a public wash-house associated with a central-heating plant. The Berlin city administration was at the same time engaged in other aspects of 'municipal socialism', with regard to public utilities and the provision of parks, sports grounds, and swimming-pools.

The achievement of the publicly supported housing associations in Weimar Berlin was truly remarkable. As is customary in Germany, the housing was not financed and built directly by the city but through the intermediary of a number of housing associations, of which the best known is the Gehag (Gemeinnützige Heimstätten Spar- und Bau A. G.). The settlements that they created gained international acclaim as did the architects who worked on them, notably Bruno Taut, but also others such as Hans Scharoun and Walter Gropius. The achievement was concentrated into the relatively short space of about eight years between the end of the inflation period in 1924 and the onset of the world depression in 1932, closely followed by the arrival of the National Socialists into power in 1933. In 1924 about 10,000 new dwellings were erected in Berlin, in 1926 about 16,000 were constructed, and in 1927 almost 27,000. Then followed a decline, to 23,000 in 1930 and 8,000 in 1933. To gauge the extent of the achievement, it may be noted that post-war West Berlin, at the peak of its building activity in 1959–60, and under much less difficult economic circumstances, built or reinstated no more than 23,000 dwellings (Kloss 1982).

6.4.4 The 'old towns' of the outer zone: Spandau and Köpenick

Spandau has the distinction of being an older settlement than Berlin. Between at latest the ninth century and the twelfth century, archaeologists have traced a sequence of Slav fortresses built on a sandy island close to the old confluence of the Spree with the Havel; from the eleventh century the accompanying civil settlement was itself fortified and had acquired an urban character. About 1180, following the conquest of the Havelland by Albrecht the Bear, the Slav location was abandoned and a new island citadel and town established in their present locations. The year 1232 is claimed, but not with certainty, as the date when urban law was granted, which would make Spandau as a town five years older than Berlin–Cölln.

The expansion of Spandau was hampered by the fortifications that in the seventeenth century replaced its medieval wall. Fortress status continued until 1903, although the fortified area was expanded in the middle of the nineteenth-century to allow some urban expansion. Few of the Prussian barracks and other military buildings have survived the combined impact of wartime bombing and the post-war construction of a ring road around the Spandau inner city; the greatest monument is the Citadel, where fortifications erected in 1560–94 after the latest Italian model contain the fourteenth-century Juliusturm.

Spandau was not only a garrison town, it was a major centre of arms production, but today's industrial skills are put to other uses. Spandau is West Berlin's most industrialized *Bezirk*. The electrical industry of Siemensstadt dominates the industrial structure; a marked post-war development has been the growth of cold-storage, warehousing, and distribution facilities serving West Berlin as a whole.

With the formation of Greater Berlin in 1920, Spandau lost its independent urban status, pride in which had been symbolized by the building in 1910–13 of yet another of the enormous town halls that are such a feature of the outer districts of the present city. However, the addition of a number of neighbouring localities produced what is now the largest West Berlin *Bezirk* by area, and the second largest in Greater Berlin. Spandau is physically rather isolated from the remainder of West Berlin by a belt of obstacles including, from north to south, Tegel Airport, Volkspark Jungfernheide, the industries, public utilities, and *Kleingärten* of the Spree valley, and the Olympic Stadium. This may well have helped the inhabitants to continue feeling 'Spandauer' rather than 'Berliner'; Spandau is the only *Bezirk* of West Berlin to have a daily newspaper of its own. In their turn, the inhabitants of the rest of Berlin have little occasion to go to visit Spandau, except for recreation in the forests and lakes that lie within its extended boundaries. It remains to be seen to what extent the recent extension of the U-Bahn to Spandau changes existing patterns of relationship.

Köpenick, among the forests and lakes of south-east Berlin, is to a large extent a mirror-image of Spandau. Old Köpenick stands on an island at the confluence of the Dahme with the Spree. At the time of the Germanic penetration into the Berlin area it was a strong point of the Sprewaner, a Slav people centred on the upper Spree valley. The Askanier established a castle on the island, which was successively replaced in the sixteenth century by a Hohenzollern hunting lodge (cf. Jagdschloss Grunewald) and then at the end of the seventeenth century by the present Köpenick *Schloss*, one of the numerous electoral palaces of the Berlin area, to which reference has already been made (section 6.2.2). It now contains East Berlin's Museum of Applied Arts (Kunstgewerbemuseum), the contents of which are very largely derived from the Kunstgewerbemuseum in the now vanished Berlin Palace.

The adjoining Old Köpenick, on its island, has had urban legal status since the thirteenth century; it has been described as still having the appearance of a small country town of the Brandenburg Mark (Pape 1985). The 1903–5 brick-gothic town hall is associated with the story of the bogus 'Hauptmann von Köpenick', the shoemaker who in 1906, relying on the instinctive Prussian obedience, put on the uniform of an officer and ordered a file of troops to follow him to the town hall, where he arrested the burgomaster and seized the treasury. Industrially, Köpenick from the middle of the nineteenth century began to take in Berlin's washing, using the clean and soft waters of the Spree as the basis of a development ranging from individual washerwomen to factory-scale establishments for steam-washing, dry cleaning, and dyeing. Köpenick resembles Spandau in extent; it is the largest of the Berlin *Bezirke*, but as much of its area consists of forests and lakes, it has the lowest population density. The combination of natural beauty and convenient situation between the rail lines eastwards to Erkner and south-eastwards to Königs Wusterhausen led before 1939 to the growth of villa colonies and weekend houses. Nevertheless, the Köpenick boundaries somewhat surprisingly include Obershöneweide which, with Niederschöneweide in *Bezirk* Treptow across the river, makes up one of the three major industrial concentrations of East Berlin. In spite of this massive industrial presence, *Bezirk* Köpenick has on environmental grounds not been selected to play any major part in East Berlin's current housing programme, which has been concentrated in Marzahn and Hellersdorf, further north.

6.5 The rural–urban fringe

The boundaries of the Greater Berlin of 1920 were generously drawn, so as to include virtually the whole of the built-up area of Berlin as it then was, together with a great deal of undeveloped land, especially on the Barnim plateau to the north-east, but also on the Teltow fringe to the south. Given the open nature of urban development outside the Wilhelmian ring, with the creation of numerous distinct 'cells' of development, it was hardly surprising that some of these cells would begin to develop beyond the city boundary, but the frontier of settlement was far from continuous. Tentacles of development emerged on the routes radiating from the city, especially as the S-Bahn pushed out suburban lines beyond the Berlin boundary. Some of these lines were electrified, for example north-east to Bernau, north to Oranienburg, and south-west to Potsdam, stimulating the growth of low-density residential development. The process of industrial decentralization also continued beyond the boundary; in particular, the Havel upstream developed as an axis of heavy industry and locomotive-building, extending through Hennigsdorf to Velten and Oranienburg. Other directions of development were east to Erkner and the cement-town of Rüdersdorf, south-east to Königs Wusterhausen, and south-west to Teltow and Babelsberg. Further out, towns such as Eberswalde or historic Brandenburg increasingly functioned as Berlin satellites (see fig. 5.2).

The largest of the peripheral settlements was Potsdam. Linked as it was to Berlin by history, by contiguity, and by its function as a commuter settlement, it nevertheless remained outside the 1920 borders. Its continuing separation reflected its distinct functions as a garrison town and administrative centre for the Prussian *Regierungs-bezirk* Potsdam.

6.6 Towards regional planning

The fragmented political structure of the Greater Berlin region before 1920, and the continuing administrative separation of city and urban field, inevitably threw up difficulties in providing for co-ordinated physical planning. Such overriding control over development as existed derived from the administrative tradition of the Prussian state, particularly the police control over building. It was this power that enabled the 1862 Hobrecht plan to cover areas outside the then administrative boundaries of Berlin (section 6.3). In this tradition also can be seen the 1875 proposal for the creation of a 'Department of the Spree' on the Parisian model. This having been rejected, the next attempt to provide regional government was the 1911 *Zweckverband von Gross-Berlin* (section 1.3). In its short life, the *Zweckverband* achieved considerable results in the co-ordination of local transport and recreational provision, but is generally regarded as a solution that came fifty years too late. Many administrative problems were then solved by the formation in 1920 of Greater Berlin, but at the price of erecting an administrative boundary excluding communities that had participated in the *Zweckverband* (section 1.4). The need for a degree of administrative co-ordination of city and urban field was recognized by the formation in 1929 of a regional planning association *Landesplanungsverband Brandenburg-Mitte*, which in 1937 was absorbed into *Landesplanungsgemeinschaft Brandenburg*; it cannot be said that either organization was particularly effective.

Urban development and redevelopment after 1945 | 7

7.1 Greater Berlin in 1945

The heavy bombing that Berlin had endured since 1943, together with the damage caused by the street fighting of the last weeks of the war, left a third of the city's dwellings totally destroyed, and many of the remainder damaged or only partially habitable. The first post-war statistics in 1946 showed that, after allowance had been made for buildings requisitioned by the occupying powers, the reduced population of the city had 8 sq. m of residential space per head, as against 16.4 in 1939; put in another way, each habitable room had on average 2.0 occupants, as compared with 1.2 formerly.

Damage was most severe at the heart of the city, where *Bezirk* Mitte and adjoining Friedrichshain and Tiergarten all lost half of their housing stock. The densely packed inner parts of Charlottenburg and Schöneberg suffered to a similar degree. In addition, nearly every public building in this central area lay in ruins: the old churches of the inner city, the Berlin Palace, the Berlin Cathedral, the Catholic Cathedral, the museums, the French and German churches and National Theatre on the Gendarmenmarkt, the Reichstag (already damaged by fire in 1933), the Kaiser-Wilhelm Memorial Church – the list seems endless. This central area of maximum devastation was referred to at the time as the 'dead eye' or 'dead heart' of the city.

There followed the great effort to rid the city of the rubble of its destroyed buildings. In 1947 34,000 people laboured at this task, many of them women, the famous *Trümmerfrauen*, who cleaned bricks and salvaged other materials that, at a time of great shortages, could be reused in the first emergency repairs of buildings. What could not be reused was heaped in great tips, the *Trümmerberge*, which have subsequently been landscaped (see section 3.2.5).

For all too short a time from May 1945, amid all the distractions of trying to patch up a great city and restore it to some sort of life, it was possible to begin to plan for the future development of Greater Berlin as a unity. The first post-war Berlin *Magistrat* had the eminent architect Hans Scharoun in charge of building matters, and he produced the first ideas on the replanning of the city. He proposed a modification of the predominant radial-concentric pattern in favour of the rebuilding of central Berlin on a linear or band principle. The Spree (to which, except in the inner city, Berlin tended to turn its back) would be followed by a belt of cultural buildings, prolonging the Prussian museums of East Berlin. South of the Tiergarten, linking with the Leipziger Strasse and the traditional Berlin inner city, would be a belt of central-area functions. North and south of these belts would be the residential districts. This linear concept was to dominate city planning in West Berlin for years to come. As late as 1955–8 West Berlin held an international competition 'Hauptstadt Berlin' (Capital City Berlin), which attracted no fewer than 151 entries, including proposals from such eminent figures as Le Corbusier and Scharoun. As if no sector boundary existed, and as if the city planning office had not been divided in 1950, the plans for the central area stretched eastwards to include the historic inner city (Pfannschmidt 1959).

The plans remained on paper only; no Le Corbusier skyscrapers were to rise above the Unter den Linden. A unified redevelopment of Greater Berlin was not to be; East Berlin had its own ideas regarding future urban development. The fact that the sector boundary between East and West Berlin ran through the 'dead heart' of the city made it all the easier for the two parts of the city to turn their backs on each other in their separate redevelopment.

7.2 The new West Berlin centre

With the creation of the 'dead heart' in the centre of Greater Berlin, the traditional locations of central functions in the urban core were abandoned, at least temporarily. A revival of retailing and other tertiary functions took place in dispersed form along the main roads radiating through the less disturbed sectors of the Wilhelmian ring and the outer zone (fig. 7.1).

One of these radials had a peculiar advantage. The Kurfürstendamm and other streets radiating from the Kaiser-Wilhelm Memorial Church in the *Zoo Viertel* had the dual advantage of having already evolved before 1939 as something of a second Berlin central area (see section 6.3.2) and of being within the 'free economy' of West Berlin. With the building of the Wall the *Zoo Viertel* began to emerge as a typically capitalist 'central business district' in a building boom that was fuelled by special tax concessions and other assistance from the federal government (Heineberg 1979a, 1979b, 1985a, 1985b, etc.). The

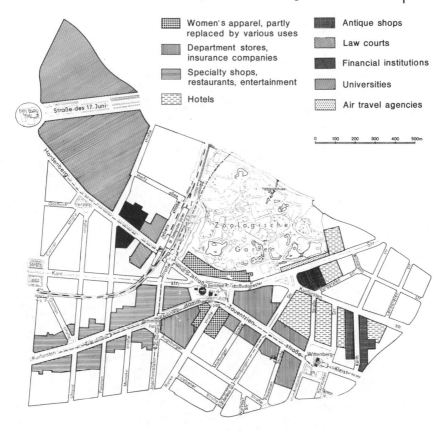

Figure 7.1 The Berlin centre (Source: Hofmeister 1980)

Kurfürstendamm and neighbouring streets evolved into one of Europe's best-known shopping areas with, in addition to exclusive clothing, shoe, jewellery, and perfumery shops, developments ranging from department stores (the restored Kaufhaus des Westens, the new Wertheim) to under-cover shopping malls (Europa Center and others).

The Zoo area also emerged as the major location of the offices of banks and insurance companies; from the early 1950s their modern office buildings provided many of the predominant architectural accents of this quarter of the city; there is a particular concentration of banks, together with the Stock Exchange, just west of the Zoo Station. In addition to its retailing functions, the *Zoo Viertel* has also West Berlin's principal concentration of high-ranking service activities of all kinds, including private medical practices, dentists, law offices,

Plate 7.1 West Berlin centre: Breitscheidplatz with Kaiser-Wilhelm Memorial Church.

accountants, management consultants, estate agents, and the like. Airline offices, national tourist offices, travel bureaus, and foreign consulates are other features. To the casual visitor the quarter is marked by the restaurants and cafés that make the Kurfürstendamm the rival of the Champs-Elysées in Paris (Wiek 1967), and are greatly frequented by tourists. Hotel provision ranges from international class (especially on the eastern side) to modest pensions in the side streets. The pre-war entertainments and cultural role have been resumed; in addition to numerous cinemas there are eight of Berlin's fifteen theatres, while the Deutsche Oper (German National Opera) is not far distant. National cultural centres are another feature, including the British Council, Amerika-Haus, and the Maison de France.

The *Zoo Viertel* has developed asymmetrically south and west of the war-damaged tower of the Kaiser-Wilhelm Memorial Church, which is its symbol and to a large extent a symbol of West Berlin as a whole. The Zoo (itself a significant tourist attraction) bars any development to the north, and academic buildings of the Art and Music Colleges and the Technical University prevent any significant extension to the north-west. This West Berlin centre is virtually unplanned, the product of unfettered private enterprise. To some extent the provision of a properly planned centre was politically impossible in the immediate post-war decades, for it would have implied acceptance that Greater Berlin was to be permanently divided, a proposition that the West Berlin government was not prepared to envisage, at least prior to the 1971 Berlin Agreement. In the early days of reconstruction after the end of the blockade the urgent necessity was to get business functioning again, repairing and adapting surviving buildings, or putting up one-storey temporary shops, following the pre-existing property divisions. The West Berlin centre evolved as and when developers were able to lay hands on building plots and see a prospect of profit. The pre-existing street patterns were retained, with the result that today massive flows of vehicles conflict with massive flows of pedestrians. Only in recent years has there been a limited rearrangement of the Breitscheidplatz, providing a traffic-free central zone linking the island site of the Memorial Church with surrounding buildings on two sides. The crowds, the disorder, the traffic, the neon signs, the outpourings of consumer goods on offer, the general commercial razzle-dazzle, all make the greatest possible contrast with the planned austerities of the East Berlin centre.

There is, however, one major difference between the new West Berlin centre and traditional city centres elsewhere; it contains no city hall (*Rathaus*) with related administrative offices. This is again because the post-war governments of West Berlin wished to do nothing to suggest that the division of Berlin might be permanent. Accordingly,

West Berlin's senate, the city's governing body, was formally established at the Schöneberg *Bezirk* town hall on John-F.-Kennedy-Platz, south of the Berlin centre. A larger group of administrative offices was established adjacent to the Wilmersdorf town hall on Fehrbelliner Platz, also south of the centre. Other offices are dispersed singly in the city (fig. 7.3).

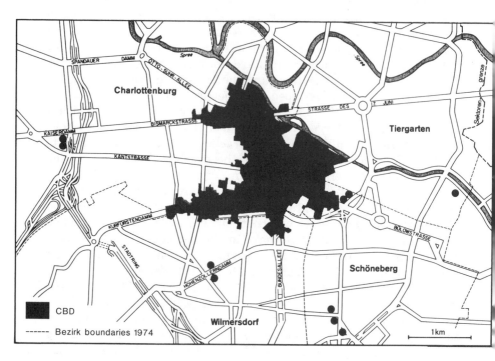

Figure 7.2 Government offices outside the West Berlin centre, *c.*1970 (Source: Hofmeister 1975)

Even less has the West Berlin centre the buildings of a national capital city; such functions have been lost. Government buildings were in any case mainly to be found in what is now East Berlin. The Reichstag building stands isolated on West Berlin soil adjacent to the Berlin Wall, on the sector boundary, and was expensively restored with the idea that the federal parliament, its various committees, and its constituent political parties would hold periodic sessions there. Such use was always infrequent and tended to incite interruptions to the access routes to the city, in support of the GDR position that West Berlin is not part of the Federal Republic. Since the Berlin agreements of 1971, such uses have ceased. 'What to do with the Reichstag building?' has become a vexed West Berlin question; in recent years it has, among other uses, been used for a standing exhibition of German history, with particular reference to the development of political institutions. The Bundeshaus, on the southern fringe of the West Berlin centre, houses the Berlin representative of the federal government, as well as representatives of the various ministries, except defence. Schloss Bellevue, on the northern fringe of the Tiergarten is held ready for the infrequent visits of the federal president. In addition the federal government has tried to compensate West Berlin for the removal of capital functions by the establishment there of a number of the less strategically important government offices, concerned, for example, with savings, insurance, or the environment (see section 5.4.4). But all this is very far removed from the significance that national administration held in the pre-war economy of the city (fig. 7.2).

7.3 Post-war housing developments in West Berlin

Efforts at housing provision in the first post-war decade took the form of repairs to bomb-damaged buildings, then of the rebuilding on bombed sites of basic five-storey walk-up apartments with individual stove-heating.

7.3.1 The *Hansaviertel*

The style of major building projects for the later 1950s and 1960s was set by the redevelopment of the devastated Hansa quarter on the western fringe of the Tiergarten. As part of the 1957 International Building Exhibition it was redesigned with contributions by such internationally famous architects as Alvar Aalto and Walter Gropius as an area zoned exclusively for residential use. Obviously, a project of this architectural quality was possible only with the generous financial assistance of the federal government. Housing predominantly took the form of high-rise apartment slabs, but they were set in generous green space, merging into the parkland of the Tiergarten, so that there was a reduction on pre-war population densities of about 30 per cent. In accordance with the ruling town-planning ideas of the time, industry and non-local commercial uses were strictly excluded; a purely residential area emerged (fig. 7.4).

Some of the residential developments of the 1950s that followed the planning principles set in the Hansaviertel were also built as reconstruction projects in the West Berlin fringe of the 'dead heart', for example the Otto-Suhr-Settlement close against the Berlin Wall in Kreuzberg. Comprehensive reconstruction of whole quarters of the city on this scale would have been impossible without 1953 legislation

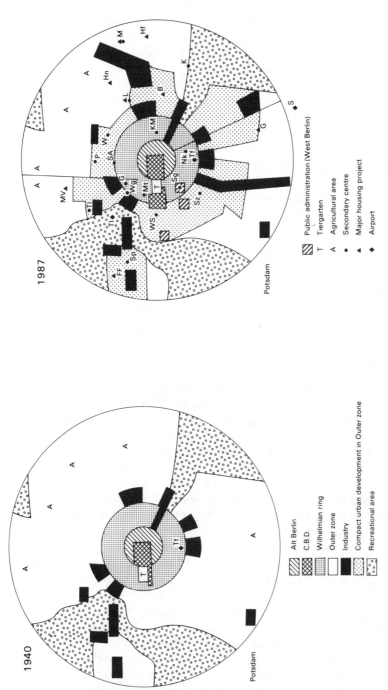

Figure 7.3 Models of urban structure 1940 and 1987. Major housing projects: (West) G: Gropiusstadt; FF: Falkenhagener Feld; MV: Märkisches Viertel; (East) B: Strasse der Befreiung/Am Tiergarten; Hf: Hellersdorf; Hn: Hohenschönhausen; L: Leninallee/Ho-Chi-Minh-Strasse; M: Marzahn. Airports: S: Schönefeld; Tf: Tempelhof; Tl: Tegel. Secondary centres: (West) G: Gesundbrunnen; Mt: Moabit; Nk: Neukölln; Sg: Schöneberg; Sp: Spandau; Sz: Steglitz; Tf: Tempelhof; Tl: Tegel; Wg: Wedding; WS: Wilmersdorfer Strasse; (East) K: Köpenick; KM: Karl-Marx-Allee; L: Leninallee; M: Marzahn; P: Pankow; SA: Schönhauser Allee; W: Weissensee.

allowing the compulsory purchase of vacant land or bomb-damaged buildings. Other developments, such as Charlottenburg-Nord, began a process of encroachment on open land in the outer zone that was to be continued in the next decade.

7.3.2 The *Grosswohnsiedlungen*

Efforts to meet the continuing housing problem in the 1960s and early 1970s involved the construction of three *Grosswohnsiedlungen*, vast housing projects on open land in the outer zone, with target populations of between 50,00 and 60,000. These were the Märkisches Viertel close to the sector boundary in the north of Berlin, Gropiusstadt (Buckow-Rudow) in West Berlin's far south-east corner, and Falken-hagener Feld west of Spandau (see fig. 7.3).

The Märkisches Viertel (the name reflects the naming of streets after towns in the Brandenburg Mark) became notorious for the problems involved in peripheral developments of this kind, problems which were not confined to Berlin. The site originally contained garden colonies that had, under conditions of housing shortage, been taken over for permanent occupation, although drainage and other facilities were quite inadequate. They had become a kind of rural-fringe slum, and the new development enabled them to be cleared. Some areas of older single-family housing were retained, providing a welcome contrast to the prevailing building style of high-rise towers and blocks. A very few tidied-up garden colonies were also allowed to remain.

To some extent the difficulties of the Märkisches Viertel were caused by changes in objectives during the planning and construction process (Hofmeister 1975a: 414–18). The number of dwellings was increased from 6,592 (plus about 800 surviving single-family houses) to 15,740 (Voll 1985b). This was in part because of the need to provide housing for persons dispossessed in the course of urban renewal programmes in the Wilhelmian ring (see section 7.3.3), but also because of the realization that, with the building of the Berlin Wall, land was finite and in short supply. The higher residential densities reflected a growing consciousness of the need to cut down the demands made for building purposes upon the limited amount of open land remaining in West Berlin. One consequence of the increase in residential units was that the shopping centre was planned on too small a scale; residents claimed that lack of competition led to high prices. There was more than the usual intensity of protest about the belated provision of facilities such as children's playgrounds, nursery schools, and paedia-tricians; as so often in such projects, those who moved in included a disproportionately high proportion of children and young people.

The plan for the Märkisches Viertel provided for two peripheral industrial areas, but these have filled up only slowly. Most of the inhabitants who are in employment have to commute to other parts of West Berlin, and access to and from this rather remote settlement continues to be a problem. The only public-transport link with the West Berlin centre is still by bus to the nearest U-Bahn station, which is hardly appropriate to a settlement of about 47,000 inhabitants. The transport situation is rather ironical in that the S-Bahn line to Frohnau actually runs beside the settlement, but until 1983 this system was operated from East Berlin, and no money was available to invest in increasing the frequency of trains or to provide a convenient station access. When the system was taken over by West Berlin in 1983, the line was even subjected to temporary closure. The U-Bahn extension to Märkisches Viertel is not expected to be completed until the 1990s. Even those with private means of transport suffer from traffic jams at peak periods. It is not surprising that a number of vociferous protest groups (*Bürgerinitiativen*) emerged to demand better conditions in the early days of the settlement.

Matters were made worse by the composition of the population, to a large extent derived either from the shanties of the former garden colonies or from the urban renewal areas of the Wilhelmian ring, where they had often been sub-tenants or had lived in emergency accommodation. Given a proper home of their own for the first time, they found the rents high and other expenses oppressive. There was also a tendency for the housing offices of West Berlin *Bezirke* to 'export' their problem families to the settlement. But after twenty years the Märkisches Viertel has become more normal, its problems no more acute than those of other outlying settlement areas in West Berlin. Demographic structure is year by year nearer to the West Berlin average, although social structure is still somewhat biased towards low-income groups. Schools and other facilities for children are now if anything in excess of requirements, and retail provision more satisfactory. Leisure facilities have been improved by the creation of a large park between the settlements and the village of Lübars. One day the U-Bahn may even arrive.

Gropiusstadt, named after the creator of its master plan, is architecturally the most distinguished of the three *Grosswohnsiedlungen*. It also has the advantage that its U-Bahn connection was completed as planned; retailing and service functions are grouped at the stations. The settlement was planned to have 16,000 dwellings and 60,000 inhabitants. From the outset Gropiusstadt had a proportion of single-family houses and apartments offered for sale, so that it is a distinctly more 'middle class' settlement than Märkisches Viertel; it is above the average in this respect both in relation to its *Bezirk* of Neukölln and to West Berlin. Its environment will be further improved with the reopening of the grounds of the 1985 Garden Show as a public park.

Falkenhagener Feld was the first of the three settlements to be initiated, intended to have 10,000 dwellings and an eventual population of 35,000. Architecturally, it is regarded as the least distinguished of the three. In its very remote location between Spandau and the border with the GDR it has severe problems of access; the promised U-Bahn has only recently reached as far as Spandau.

Coupled with the decline in the residential population of the West Berlin centre and the impact of urban renewal in the Wilhelmian ring, the *Grosswohnsiedlungen* contributed to a marked centrifugal movement of the West Berlin population and a general increase in the settlement density of the outer zone (further outward movement being prohibited by the existence of the GDR boundary). The settlements were visited by admiring architects from all over the world, but today they have fallen almost totally out of favour in town-planning circles. High-rise buildings, exclusive residential zoning, one-class residential areas, and long journeys to work from peripheral locations are all regarded as socially undesirable. In more specifically Berlin terms, the taking of any more of the scarce resources of open land for building purposes is now regarded as unacceptable. The period from the mid-1970s was to see West Berlin building operations taking very different directions.

7.3.3 Urban renewal

When the rebuilding of some totally destroyed inner areas, such as the Hansaviertel, was drawing to its close in the late 1950s, attention turned as a second priority to the problems of areas that had largely escaped destruction, but where the building fabric was outworn and considered to be unsatisfactory as a human environment. Essentially, we are concerned here with a narrow belt between the southern boundary of *Bezirk* Mitte (now marked by the Berlin Wall) and the former Customs Wall (southern Luisenstadt) together with the very much larger area provided by the West Berlin section of the Wilhelmian ring. There is no doubt about the poor condition of these areas. As noted in section 6.3.1, the *Mietskasernen* of these areas were characterized by extremely high residential densities, a maze of dark internal courts and a predominance of tiny dwellings, often consisting of only one room and a kitchen. An area of 107 hectares at Kreuzberg-Kottbusser Tor, surveyed in 1962 prior to renewal, had 37,000 inhabitants. Of its 16,000 dwellings, 91 per cent had been erected before 1918, of which 75 per cent dated from before 1870 and 40 per cent faced onto internal courts. Only 15 per cent had an internal WC and either a bath or shower; another 19 per cent had an internal WC only, while 66 per cent had to use toilets on the staircases (Pirch 1985).

Heating was by individual stoves, the brown-coal briquettes used contributing to the characteristically high air pollution of these inner areas of the city. In addition, the buildings had all sorts of constructional defects, from rising damp to rotting windows and leaking roofs. There was a general lack of public gardens, children's playgrounds, and sports facilities.

Another problem perceived at the time of survey was the intermixture of residential and other uses; the area mentioned above was found to include 1,740 businesses employing 16,300 people. Half of them were concerned with local retail distribution, mostly very small shops, but the remainder included numerous manufacturing and craft establishments, typically located in the internal courts. This intermixture of residence and economic activity had its advantages in terms of employment provision, reducing the time of the journey to work for many inhabitants, but obviously manufacturing activities could be a source of noise disturbance, atmospheric pollution, and traffic generation. The town-planning wisdom of the time was to regard such mixed uses as undesirable, to be eliminated if at all possible. There were also symptoms of economic weakness. The areas deemed to be in need of renewal wrapped round *Bezirk* Mitte in the heart of Berlin, with which they were formerly closely connected. The clothing workshops, printing works, and furniture manufacturers of the southern Friedrichstadt and southern Luisenstadt, for example, worked for customers in what is now East Berlin. All the main lines of movement, for business, work, retailing, and recreation ran in that direction. Now all links are sundered by the Berlin Wall, and some of these areas, notably Kreuzberg *Bezirk*, but also Tiergarten and Wedding, suffer from a distinct degree of isolation from the rest of West Berlin. A certain amount of dereliction and redundancy of buildings can be detected as a result.

A complication in the urban renewal process was provided by the demographic and social characteristics of the resident population concerned. In a West Berlin that had above-average numbers of elderly inhabitants, especially elderly women living alone, these areas were above even the West Berlin average in this respect. It was to a certain degree a relict population; apart from the men who had failed to return from the war, the younger and more enterprising element had tended to move away to more attractive areas of West Berlin. The old remained, with their accustomed neighbours, their accustomed streets, and their accustomed small shops. They were joined by other economically weak groups, prepared to accept inferior accommodation in return for low rents: the chronically sick, the chronically unemployed, the single-parent families. Many such resided as sub-tenants. They were joined from the 1960s by a very different group, the foreign workers and their large families, especially the Turks, who rapidly became the predominant element in the populations in parts of the Wilhelmian ring, notably northern Kreuzberg. The members of these groups were unable to contribute financially to any renewal scheme, and unwilling to move away to more expensive if superior accommodation elsewhere.

Faced with the often appalling condition of buildings and the apparently inextricable mixture of uses, the first solution adopted involved the complete clearance of the areas concerned, followed by rebuilding. This approach had already been adopted in two areas of Wedding in the 1950s. It was particularly attractive where major road schemes were in prospect, as along the southern side of *Bezirk* Mitte through Kreuzberg. The Kreuzberg-Kottbusser Tor clearance area had already lost 81 per cent of its fabric, together with 40 per cent of its inhabitants and 70 per cent of its businesses, before a major road project there was abandoned in 1976 (Pirch 1985).

By the later 1960s both the results of surveys and questions of cost had persuaded the West Berlin senate that a more flexible approach was essential. The existing street patterns were to be retained, the better buildings along the streets modernized, the interiors of blocks 'de-cored' of their outbuildings and equipped with gardens and games pitches, but not all of the existing businesses driven out. Even this modified programme ran into opposition. Complete modernization to the same standard as in new buildings was extremely expensive and, even without attempts to cash in by private landlords, necessarily involved increased rents. The process was also felt to be excessively paternalistic and bureaucratic; modernization was imposed irrespective of the opinions of the occupants, a third of whom had to depart anyway because of the lower residential densities achieved in the process. The whole business became politicized. Groups of what might loosely be described as students occupied apartment houses due for renewal, aiming to show that much more modest measures undertaken in close consultation with the existing residents could provide entirely acceptable results without the need for vastly increased rents. The occupied houses were immediately apparent because of the often bizarre painting of their façades and the numerous banners hanging from their windows. Attempts by the senate to repossess them using police led to pitched battles in the streets in 1979–81. Yet it was in the end the senate that changed. The surviving occupations were licensed to continue (some just wasted away when the senate ceased to attack them). From 1981 the senate adopted the policy of 'repairs before modernization, and modernization before urban renewal' (Pirch 1985).

The renewal process was also affected by the decision that the next Berlin International Building Exhibition (IBA) in 1987 would not complete one prestige project, like the Hansa Viertel, but a number of separate projects scattered throughout West Berlin. Some of these were to be in the field of urban renewal, and were to be particularly centred on Kreuzberg. The IBA principle was opposed to the earlier processes of the 1960s and 1970s, which combined high financial expenditure with maximum disruption of community. The emphasis was on a modest stage-by-stage improvement of living conditions, likely to appeal to existing inhabitants and to be cost-effective, with the ultimate objective of stabilizing the areas concerned both residentially and as locations for economic activities. Simultaneously there was a general up-grading of the residential environment, with the provision of small public gardens and games spaces.

7.4 *Cityband* and Wall fringe

As already noted (section 7.1), the first plan for the rebuilding of Berlin involved the notion of a linear or band structure for the central area, embracing the central parts of both West and East Berlin. The definition of the *Cityband* remained far from fixed. Some saw a differentiation between a cultural band running along the Spree north of the Tiergarten and a business–administrative band to the south, but others have stressed the need to treat the whole of the Tiergarten area as one. Similarly, some have defined the band widely so as to include the whole of the present centres of West and of East Berlin, including the potential extension of the latter into the southern Friedrichstadt, in West Berlin territory. Others have seen it in a narrower sense as a concept essentially concerned with the planning of the space surrounding the Tiergarten, lying between the two existing centres (Hofmeister 1975a: 314–21). What is clear is that after the break-up of any joint planning for Berlin in 1950, and even after the building of the Wall in 1961, the West Berlin administration was careful to ensure that any major developments in the area between the two centres, or immediately south of the East Berlin centre, would still be viable in the event of a reunification of the city, in which the two centres would function as one. Similarly, nothing should be done that would stand in the way of such a merging of the two centres.

The principal *Cityband* developments on West Berlin territory have been cultural in nature (fig. 7.4). The small 'Berlin-Pavillon', created at the time of the 1957 Building Exhibition adjoining the Hansa Viertel, serves as an information point on building developments in Berlin, with special exhibitions from time to time. The 1960 Academy of Arts resulted from private American generosity, while the Congress Hall was the contribution of the United States to the 1957 Building Exhibition. (The great parabolic roof of the main hall collapsed without warning in 1980, but it is likely that, after restoration, the building will continue to contribute to West Berlin's life.) The northern Tiergarten belt ends eastwards at the Reichstag building, which, as has been mentioned, has yet to find a post-war purpose. Some would like to create here a museum of German history, others propose that the open spaces of the Platz der Republik in front of the Reichstag be filled with buildings containing the various official buildings of the Federal Republic that are at present scattered throughout West Berlin. Then, within a few metres of the windows of the Reichstag, is the Berlin Wall.

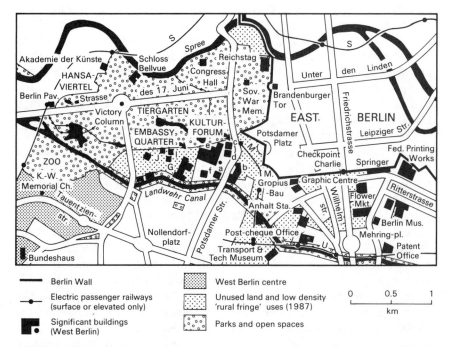

Figure 7.4 *Cityband* and Wall fringe. Buildings of Kulturforum: a: National Gallery; b: State Library; c: Philharmonie (concert hall); d: Chamber-Music Hall; e: Museum of Western Art (under construction 1987; Museum of Applied Arts complete); f: International Science Centre. Surviving embassy buildings: I: Italy; J: Japan; G: Greece. Railway systems: U: U-Bahn; S: S-Bahn; M: Experimental magnetic railway.

South-east of the Tiergarten is the new Kulturforum. It lies in an area that Hitler's town planners and wartime destruction left almost devoid of buildings. Only the nineteenth-century Matthäus Church remains on a Kemperplatz that is now surrounded by a group of cultural buildings waiting to be reunited with the historic core of East

Berlin beyond the Berlin Wall. The buildings include the National Gallery, the Kunstgewerbemuseum (Museum of Applied Arts), which is the first stage of a complex to unite on this site all medieval and later western art, the Philharmonie (concert hall) and associated Chamber Music Hall and, rearing up its giant book-stack against the Wall, the State Library. An adjacent nineteenth-century government building has also been restored and expanded as a centre for scientific and academic exchange.

However worthy the insistence on looking towards a reunited Berlin, it has to be accepted that these buildings are extremely isolated within the present structure of West Berlin, being situated far to the east against the Wall and further cut off by the Tiergarten. There are no passing crowds, nobody 'drops in' on impulse. There are no shops, no restaurants. People come by bus or private transport specifically to attend a concert, to visit a museum, or to read in the state library, then go away again: there is no street life, no 'life of the quarter', if one excepts the small boys who come to roller-skate on the podium of the National Gallery. Facilities which could have enlivened other parts of Berlin here lie in a dead area.

The southern fringe of the Tiergarten was earmarked by Hitler as a special quarter for embassies, many of which were already established in villas here. Today the area mainly lies waste, its future a planning problem. Certainly, its immediate future cannot be with embassies, since West Berlin has lost its capital functions. The vast and semi-ruined Italian Embassy of the National Socialist period serves in part for consular purposes, the adjoining Japanese Embassy of the same period is to be restored as a German–Japanese cultural centre. Remoteness from the West Berlin centre, the fact that so much of the land belongs to foreign governments, and thoughts about possible usage in a reunited Berlin, have so far inhibited redevelopment; the laying-out of a park in extension of the Tiergarten seems to be the most likely fate of this land, at least provisionally.

The area along the West Berlin side of the Berlin Wall presents an incredibly negative spectacle. The land lies bare and unused, or is devoted to uses that in a normal city are characteristic of the 'urban–rural fringe'. In part this reflects the policy of the Berlin senate, already referred to above, to allow no building that would stand in the way of an eventual linking of the West Berlin and East Berlin centres in the context of a reunited Berlin. It also reflects the way in which West Berlin has turned its back on the area along the Wall; what was once the heart of a city of world status becomes an extraordinary version of a rural fringe, an extraordinary inner-urban 'Wall fringe'.

The process of fringe development becomes apparent quite rapidly on moving east from the West Berlin centre; luxury hotels, office blocks and good-quality modern apartments give place to uses requiring a large amount of space, such as furniture warehouses and their parking lots, and used-car lots, which would normally be found along the roads leading out of a city. Perhaps the gambling joints, the associated sleazy cafés offering 'breakfast 24 hours', the ladies of easy virtue and the (alleged) drug trafficking of the northern Potsdamer Strasse could also be regarded as a fringe phenomenon peculiar to West Berlin.

The section of the Wall fringe from the Potsdamer Platz (once the bustling 'Piccadilly Circus' of Berlin and now within the Berlin Wall system of obstacles) to the southern Friedrichstrasse is an extraordinary spectacle. One block near the Potsdamer Platz is reserved for the training of guard dogs, another is the permanent headquarters of a circus, yet another muddy area is the scene of a regular Saturday market; to some extent this land on the approaches to the former Potsdam Station has remained unoccupied because, until recently, it has belonged to the Deutsche Reichsbahn, which is based in East Berlin (see below). This area is now traversed by an experimental overhead line for a transit system worked magnetically, which may improve access from the U-Bahn to the Kulturforum.

Moving eastwards round the southern fringe of the East Berlin *Bezirk* Mitte, the southern Wilhelmstrasse, instead of looking like the 'Whitehall of Berlin', looks like a tree-lined country road. Frontage uses, at least until recently, included a track for learning to drive and dumps for gravel and other building materials. The northern end of the street, where it is blocked by the blank face of the Wall, was for long marked by a ramshackle pub appropriately called by the English name of 'Land's End'. A little further east the southern Friedrichstrasse has garden centres (no doubt in part reflecting the proximity of the West Berlin wholesale flower market). Further east still is a kind of farm, with facilities for riding ponies. Yet all these 'rural fringe' uses are on land that was once in the very heart of the city.

Among the few surviving buildings of the southern Friedrichstadt, right against the Wall, is the Martin-Gropius-Bau, a nineteenth-century museum building originally intended as a Museum of Applied Arts (later moved to the Berlin Palace, and later still divided between the East Berlin museum at Schloss Köpenick and the West Berlin museum at the Kulturforum). It is now the proposed location of the Museum of German History. Beside it is a piece of German history that many would like to forget, the site of the SS and Gestapo headquarters and torture chambers in the former Prinz-Albrecht-Palais. For years the site has lain buried under heaps of building materials; now there is a search for an appropriate memorial to the unspeakable.

With the completion of the 1971 Berlin Agreement it became apparent that the reunification of Berlin was to be deferred indefinitely, and West Berlin's attitude to the *Cityband* concept was modified accordingly. In particular, it was no longer held to be necessary to hold open the southern Friedrichstadt for a later expansion of the administrative quarter around the Wilhelmstrasse. Accordingly, a number of rebuilding schemes have been redeveloped here as demonstration projects in connection with the 1987 International Building Exhibition. They are characterized by low-rise buildings, either following existing street lines or built round internal garden courts, with a mixture of residential, retailing, office, and even selected industrial uses.

The same principles are now applied to rebuilding in West Berlin generally; the time of vast high-rise peripheral developments is past; the emphasis is now on more modest schemes, filling in the remaining vacant areas in the inner parts of the city.

The general dereliction of the Wall fringe south of *Bezirk* Mitte is increased by the problem of the disused railway land on the approaches to the functionless Potsdam and Anhalt Stations. The Potsdam Station itself lay within East Berlin and has totally disappeared; of the vast Anhalt Station only a fragment of the façade and entrance portico remain. For many years the railway land lay unused and unusable, because it belonged to the GDR-based main railway system (Deutsche Reichsbahn). It is now in the possession of the West Berlin government and is likely to be used for the creation of parks and recreational facilities, which are severely deficient in this part of Berlin. Some buildings on the approach to Anhalt Station have been transformed into a Museum of Transport and Technology (the earlier Transport Museum in the buildings of the former Hamburg Station in the Wall fringe further north has also been returned by the GDR to full West Berlin control).

7.5 Secondary centres in West Berlin

It has already been noted that the wartime destruction of the old central retailing district centred on the Leipziger Strasse led to a dispersal of service functions to enlarged or even new secondary centres in both East and West Berlin; one of these evolved into the West Berlin centre in the *Zoo Viertel*, the primary centre serving the whole of West Berlin (see section 7.2).

These secondary centres were characteristically linear in form, extending along the main roads radiating from the centre through the Wilhelmian ring. They often coincided with old village nuclei, as for example Schöneberg and Steglitz on the road to Potsdam. 'Thickening' of the centres through retail development into the side-streets is rare. The larger secondary centres in particular provide for far more than daily needs; their department and chain stores, as well as specialist shops, compete in fields like clothing, shoes, and household goods with stores in the West Berlin centre. Although hinterlands overlap, this type of retail provision has been found to attract purchasers from a radius of over 4 km (Aust 1970). The attraction of certain centres has been increased not only by the building of large new stores but by public transport improvements, such as the U-Bahn extension to Steglitz.

Three of the secondary centres exceed all the others in importance and have the greatest degree of growth, although suffering to some degree from the competition of the West Berlin centre. They are the Schlossstrasse in Steglitz, the Wilmersdorfer Strasse in Charlottenburg and the Karl-Marx-Strasse and Hermannplatz in Neukölln. The success of the first two is hardly surprising, coinciding as they do with Berlin's west and south-west sector of higher-income development.

The success of the Steglitz secondary centre has been examined by Heineberg (1985b). The centre gained an initial impulse from the lesser degree of wartime damage to both its retailing facilities and its tributary residential areas, as compared with the traditional retailing centre on the Leipziger Strasse and with the Charlottenburg and Zoo areas of West Berlin. The centre gained from being on the main south-west radial road linking central Berlin with Potsdam (at least until cut by the Wall at each end), and was able to take over some activities formerly located in the Leipziger Strasse. Transport advantages were subsequently accentuated by proximity to the terminus of one of the access roads to the inner-ring autobahn and even more by the construction of the U-Bahn. The revival of the S-Bahn system will presumably be a further advantage. In contrast with secondary centres in East Berlin, but in common with all West Berlin centres, the rapid expansion of Steglitz reflected the prevailing free-market ideology and the benefit of special federal government support to investment in West Berlin, in the form of tax concessions and other assistance.

This shopping centre, that before 1939 had not a single department store, acquired no fewer than three, as well as two large low-price chain stores and branches of four of the large clothing chains. There were also numerous specialist shops, especially for clothing and shoes. They were joined in 1970 by Forum Steglitz, a totally enclosed project with five sales floors, containing a food supermarket and up to 130 retailing units (the size and number of individual units tend to fluctuate rapidly according to their commercial success or otherwise). The building also puts under cover the stalls of the popular Steglitz street-market. A more dubious development is the Steglitzer Kreisel, in a somewhat isolated position at the southern end of the centre. This seventeen-storey office block, with retailing facilities at its base and built over a U-Bahn station, a bus station, and parking facilities, has remained a 'white elephant', one of the most prominent of the building scandals to which West Berlin is unfortunately prone. The costs of this private but publicly supported project having gone vastly over estimate, it has absorbed millions of marks provided by the tax-payers of the Federal Republic and of Berlin. No commercial tenant having been found, the building is largely occupied by various offices of *Bezirk* Steglitz. The opening of a large hotel at the foot of the tower has done something to brighten-up the scene (Heineberg 1985b).

The second most important of the West Berlin secondary centres is exceptional in that it is not on one of the radials, although the Wilmersdorfer Strasse in Charlottenburg is an early road, that today acts as a link between streets radiating from the Zoo area, notably the Kantstrasse and Bismarckstrasse, which are themselves lined, at least intermittently, with small shops and restaurants. It was already of some importance as a shopping centre before 1945, benefiting from proximity to the Charlottenburg main-line and S-Bahn station, then more important than it is today. After considerable wartime destruction, the Wilmersdorfer Strasse has acquired the customary assemblage of department and chain stores, but not quite on the scale of Steglitz. The centre has gained from the construction of two new U-Bahn stations, and as it is not a main through-traffic street it has been possible to pedestrianize a considerable portion. Irrespective of commercial advantages, the new street landscape is extremely pleasant, with raised flower beds and kiosks for selling ice-cream or hot sausages, according to season. When the weather is good the seats are filled with elderly persons, who obviously enjoy watching the world go by.

Of the lesser secondary centres, mention should be made of Badstrasse/Brunnenstrasse in Wedding, situated on a radial that continues the Rosenthaler Strasse in East Berlin. In the early post-war years the gaps caused by bombing were filled in with booths and simple one-storey shops, and the centre developed as something of a 'frontier market', thanks to its proximity to East Berlin (surrounding it on three sides) and to Gesundbrunnen, the first station within West Berlin on the northern section of the S-Bahn ring. Everything changed in 1961 with the creation of the Berlin Wall and the division of the S-Bahn system. Although the shops were improved physically in the course of urban renewal programmes, the local working-class and foreign-worker populations had insufficient purchasing power to prevent the absolute and relative decline of this centre, which is characterized by retail and service provision of somewhat low standing (Heineberg 1985b).

7.6 Urban redevelopment in East Berlin: the early years

As in West Berlin, the early post-war years were ones of rubble-clearance and basic repair-work with salvaged materials on any damaged buildings that could be made habitable. The 1946 Conference of the SED had already called for a planned development of housing and the preparation of reconstruction plans, but there was effectively no construction of new housing in the GDR before 1950 (Lammert *et al.* 1969).

The year 1950 was in a number of ways a turning-point. Legal changes meant that property and buildings could be acquired by the state when needed for construction projects. The ownership and price of land ceased to be a consideration in urban development. Under the type of socialist system operated in the GDR, the optimum use of land is not what gives the greatest return in a free market but what the state believes to be of the greatest benefit.

1950 was also the first year of a five-year plan in which housing construction was given a firm if modest place. In the same year, the government issued 'sixteen fundamental principles of urban development', which were largely incorporated in the 1950 'Law on the building of cities in the GDR and of the German capital city, Berlin' (*Aufbaugesetz*). Among other points, it was stressed that town planners had to maintain a close awareness of the relationship of urban planning to the needs of industrial development. Planning, however, had not to be purely utilitarian but had to be aesthetically pleasing and express the new social order of society. In practice this seems to have meant that buildings of high social significance, such as a House of USSR–GDR Friendship, a youth centre or an opera house would become urban dominants, rather than banks or department stores. Wide radials along which organized processions could converge on a central square designed for mass demonstrations were also a necessary feature.

When rebuilding recommenced on a small scale in 1950, it took the form, as in West Berlin, of simple three- or five-storey walk-up apartments built in devastated industrial quarters, such as Friedrichshain. The main feature of the 1950s, however, was the extraordinary decision, in a time of continuing shortage of labour and materials, to concentrate resources not even on the East Berlin centre but on a single massive development along the Grosse Frankfurter Strasse/Frankfurter Allee, the major radial leading eastwards out of the inner city. The development thus falls in line with the tendency for the secondary centres of post-war Berlin to develop on the radials through the Wilhelmian belt (see section 7.5), but the rebuilding of the Stalinallee, as it was renamed (later again renamed Karl-Marx-Allee), could scarcely have been more different from, say, the commercial transformation of the Schlossstrasse in Steglitz.

Basically, this was a parade street, between 75 m and 80 m wide, such as might have been created in the capital of an eighteenth-century principality, but this was 'the first socialist street in Berlin'. It was lined on each side with a continuous curtain of seven- to nine-storey apartments, built by conventional techniques but with façades clad with classical motifs and decorative tiles (which as in all modern buildings have a distressing tendency to fall off). Major intersections were marked by elaborate towers, fountains, and statuary. A benevolent statue of Stalin (subsequently removed) presided over the scheme. This elaborate 'Stalin' style of architecture would be familiar to the inhabitants of almost every major city between Berlin and Vladivostok. At street level there were shops and restaurants 'restoring again the life of a great city to Berlin' (it must be remembered that at this stage the rebuilding of the East Berlin centre had not yet commenced, so that the Stalinallee had a greater functional significance than at the present day).

One of the objectives of the scheme was to demonstrate a new attitude to housing provision, by building in the working-class and hitherto rather neglected district of Friedrichshain, housing of the highest standards, to be allocated on the basis of social equality and social need, not of wealth. The expensive ornamentation of the buildings also proclaimed that henceforth nothing was to be too good for the working class. In contrast to the adjoining *Mietskasernen* the new apartments had no internal courts, were light and sunny, and were equipped with all modern amenities, such as an internal WC, bath, hot water supply, and central heating.

In its spatial relationships, the Stalinallee was to provide the same sort of triumphal axis to the east that the Unter den Linden axis had long provided to the west, thus organically linking the working-class quarters with the centre of the city (Lammert *et al.* 1969). The location was well chosen with regard to public transport, having a direct U-Bahn connection with the core of the city. The project was declared to be of national importance so that workers from all parts of the GDR as well as thousands of volunteers contributed to the construction work.

Another remarkable achievement of the 1950s, again in spite of the acute shortage of materials, was the restoration of the historic buildings along the Unter den Linden, among them the Brandenburg Gate, the State Opera, St Hedwig's Catholic Cathedral, the various buildings of the Humboldt University, the seventeenth-century Zeughaus of Andreas Schlüter and others (Armoury, now Museum of German History) and Schinkel's 1816–18 Neue Wache (Guard House), now 'Memorial to the victims of fascism and militarism', before which the goose-stepping armed guards attract the tourist crowds. At the same time the vast and ornate Soviet Embassy was rebuilt, occupying the area of the old Imperial Russian Embassy and much more land besides. The British and French Embassies, nearer to the Wall, have not been rebuilt.

The record of the GDR government in the restoration of its damaged historical monuments is in general extremely good. It is only necessary to think of the care lavished on the palaces and other buildings of the park at Potsdam, the years of work on the two great churches and the National Theatre (now concert hall) on the Gendarmenmarkt (Platz der Akademie), or the restoration of Berlin's massively ornate Protestant Cathedral, which many would have been happy to see consigned to oblivion. There are a few black marks, among them the destruction of the damaged town palaces of Potsdam and Berlin. The Berlin Palace was considered by many experts to be repairable, and worth saving because of its historical and architectural importance. Its destruction appears to have been partly because it was perceived as a symbol of an oppressive past, at a period when the GDR was anxious to stress its new beginnings, rather than, as in recent years, seeking legitimacy by stressing its links with German history. Another reason appears to have been sheer expediency, the wish to create on the site of the Berlin Palace and the adjoining Lustgarten a Marx-Engels-Platz vast enough to contain the mass demonstrations that punctuate the Communist year. For the rest, most of the old inner city lay empty through the 1950s, the rubble cleared, its grass-grown expanses interrupted only by the restored Marienkirche and town hall.

7.7 Prefabricated housing

By the middle of the 1950s it was apparent that if the traditional building methods used in the Stalinallee were to continue to be used, then the expansion of housing provision that was so desperately needed in Berlin and elsewhere could not be achieved, because the demands for labour, especially for skilled building craftsmen, could not be met. Attention then turned to the use of the prefabrication techniques that were also being developed in the Soviet Union and other Soviet-style socialist countries. At first prefabricated elements were used only to build the shell of the building, leaving the interior finishing to be done by hand. Later the system of putting buildings together from *Grossplatten*, concrete wall slabs that could make up the whole external wall of an apartment, with corresponding wall and floor slabs, removed the distinction between shell and interior. Much of the building work was transferred to the special plants making the concrete sections; the actual building work became more of an assembly operation, rather as in the mass production of vehicles.

The introduction of prefabrication did not simply involve the use of pre-cast units; a whole support organization was involved. Plants had to be set up for casting the units, and special road transporters provided to move them to the building sites. Hoists for loading and unloading were needed, and tower-cranes for putting the sections in place.

The results at first were, admittedly, rather monotonous. The narrow range of units available could be combined into only a limited range of apartment plans, varying only with the size of family; this indeed is still true today. The uniform apartments were then assembled in combinations determined by 'crane ideology', that is to say in long, straight slabs, all running in the same direction, beside which the cranes could run on rails. Here at least, time has brought progress, the ingenuity of architects has been used to produce more imaginative combinations of the inflexible units. Some of the first breakthroughs were made at a very early stage, in the beginning of work in 1958–9 on the extension inwards to Alexanderplatz of the rebuilt Stalinallee (from 1961 Karl-Marx-Allee). Here the quantitative leap was made from five-storey blocks to ten-storey blocks with lift-service. The residential blocks were no longer lined up in corridor style but stood back on building lines 125 m apart, with shops, cinemas, and restaurants standing forward from them 'to provide a new type of street scene' (Lammert et al. 1969). Subsequent critics were to regard this linear dispersion of functions as 'lacking in urbanity', as failing to generate the bustle and interest of a true city centre.

The other innovation that arrived in the 1960s along with prefabrication was the notion that it was not in fact necessary to line up residential buildings along a street; they could be developed in depth, as complete settlements with apartment blocks standing in plentiful green space (a layout in any case suited to 'crane ideology'). The necessary retail and social provision could then be provided in traffic-free conditions within the settlement, not on the bounding streets.

The first of these new-style developments were built in the 1960s at locations relatively close to the East Berlin centre, for example north and south of the Karl-Marx-Allee. Contrary to developments in capitalist cities, the East Berlin authorities tried in this way to counter the drift of population from the centre; residential accommodation even 'bridged over' the East Berlin centre using the higher storeys that in a capitalist city would certainly have been devoted to office uses (see section 7.8).

A more decisive change in the urban structure came in the 1970s, when it was accepted that the acute East Berlin housing problem could not be solved without major urban extensions. As in West Berlin, these took the form of peripheral *Grosswohnsiedlungen* designed for up to 100,000 population each. They were predominantly concentrated in a sector east of the S-Bahn ring, that had remained largely undeveloped until 1945. The first really large development was at Leninallee/Ho-Chi-Minh-Strasse, immediately outside the S-Bahn ring on the axis of the Leninallee (Landsberger-Chaussee). More recent major developments were further out, the greatest Marzahn, intended for 160,000 inhabitants. This settlement had the planning advantage of proximity to the principal East Berlin area of new industrial development in Lichtenberg-Marzahn (Zimm and Bräuniger 1984), and is served by a number of stations on the electrified S-Bahn line to Ahrensfelde. Other major developments in this sector include Hohenschönhausen (also with S-Bahn service), with a target population of 100,000, and Am Tierpark/Kaulsdorf/Hellersdorf, which are being connected by an extension of the U-Bahn rather than the overloaded S-Bahn system (see fig. 7.3).

A characteristic of these new settlements is the simultaneous provision of the necessary social equipment. Each building development is automatically accompanied by a capital credit, amounting to 50–60 per cent of total cost, which is devoted to the technical infrastructure of roads, sewers, and the like, but also to social provision. The new developments are accompanied by neighbourhood facilities, such as crèches, kindergartens, health centres, schools, basic supermarkets, cafés, and the characteristic GDR combined *Dienstleistungen* establishments for all sorts of services and repairs. Major settlements, such as the Leninallee, also have large central shopping centres that rank as major secondary centres on the scale of the city (see section 7.10). As with new peripheral settlements everywhere, there is a problem that the initial population has an unusually high proportion of children, which subsequently disappears as the population structure becomes normal. To avoid the creation of unnecessary facilities, crèches and kindergartens are created out of the ground floors of apartment buildings, which are subsequently converted into normal apartments, especially for elderly persons who prefer ground-floor locations. Heating and hot water are provided from district heating plants, which are increasingly taking over this function throughout East Berlin.

According to a statement by First Secretary Erich Honecker in 1986, between 1971 and 1985, 213,350 dwellings in East Berlin were rebuilt or modernized, improving the housing conditions of 600,000 Berliners, over 50 per cent of the capital's inhabitants. In spite of these improvements, rents have not changed for years, amounting to not more than 5 per cent of the family income.

Plate 7.2 Marzahn housing project: waiting for the Sunday afternoon 'youth dance' at a neighbourhood restaurant and bar. The residential blocks are are treated from large prefabricated enamel panels

7.8 The East Berlin centre

The planned development of the East Berlin centre (plate 7.3) began only in the late 1960s, very much later than the more spontaneous, capitalistic development of the West Berlin centre (Heineberg 1979a, 1979b, 1985a, 1985b). The plan involved the creation of a substantially traffic-free area taking up the greater part of what had been the historic inner city, an area which, before 1945, had become a somewhat run-down fringe of the total urban core. In an area that the war (and the first post-war GDR leader Walther Ulbricht) had left almost totally clear of buildings, the planners created a pattern of four great squares (see fig. 6.1).

In the south-west the Unter den Linden leads across the Marx-Engels-Brücke (formerly Schlossbrücke), its marble statues returned from West Berlin, into the Marx-Engels-Platz. This 'government' square was created out of the site of the former Berlin Palace and the former Lustgarten, to be the location of mass demonstrations on 1 May and other dates in the Communist calendar. It is overlooked by the new buildings of the Staatsrat (Council of State) and the Foreign Ministry, as well as the historic Zeughaus (Armoury) at the head of the Unter den Linden, the Altes Museum, and the Protestant Cathedral. It seems likely that the present historically minded leaders of the GDR would have been happy to receive ambassadors and distinguished guests in a rebuilt Berlin Palace, had it been allowed to survive, but this was not to be. Instead a large part of the square is now taken up by the 1976 Palast der Republik, a modern building curiously combining the functions of a chamber for the GDR parliament and an entertainments centre. The mass demonstrations are no longer held here; the remainder of the square is used as a car park for bureaucrats, which seems to be an unworthy use of such a key location.

The next square to the north-east, beyond the Spree, is the Marx-Engels Forum (plate 7.4). It centres on statues of a seated Karl Marx, with Friedrich Engels standing behind him (Berlin joke – Marx: How much has all this cost? Engels: You had better sit down before I tell you). The Forum is bounded to the south-east by the Nicolai quarter, where exceptionally a few pre-war houses survived, and where the Nicolai church (the oldest foundation in Berlin) and the buildings of the quarter have recently been sympathetically restored. High-priced cafés and boutiques front onto the Forum. The delivery of dismantled façade elements from West Berlin has facilitated the reconstruction of the 1776 merchant's house known as the Ephraimsche Palais.

The very large square further to the north-east is dominated by the East Berlin TV tower, a major tourist attraction; it also contains the restored Marienkirche, second oldest foundation and oldest surviving

Plate 7.3 Rebuilding of the East Berlin centre. The massive development of prefabricated high-rise blocks characteristic of the 1960s and 1970s (in the background, beyond the 'Red Town Hall') was replaced in the 1980s by an attempt at more human-scale development of the Nicolai Quarter, with high-priced restaurants and boutiques. Note that the upper floors, even in the centre of Berlin, are residentially occupied, rather than used as offices as in city centres of capitalist society.

Plate 7.4 The Marx-Engels monument and Berlin cathedral.

church building in Berlin, and is overlooked by the 1861–70 red-brick Rathaus (town hall). The square is bounded east and west by two levels of specialist shops, restaurants, and tourist offices of socialist countries, above which are residential apartments. The covered retail market reflects a centuries-old market tradition.

Finally, beyond the elevated S-Bahn line marking the former wall of the inner city, is the enlarged Alexanderplatz. This gateway to the Eastern quarters of Berlin is a traditional centre of 'popular' life. Today it contains the large Centrum department store and a hotel tower, and is fringed by specialist shops. It is overlooked from the north by the tower-block housing the offices of the state tourist organization and by the State Police headquarters.

By comparison with the 'spontaneous', disorderly growth of the West Berlin centre, with its emphasis on retailing and office development, the East Berlin centre is characterized by order, planning, and the representative functions of a capital city, albeit one controlling a very much smaller area than was governed from the city before 1945. The limited shopping floorage is (apart from the covered market and the single department store) very much taken up as display space for leading GDR firms, as show-places for the various Soviet-style socialist states, and for specialist shops selling books and gramophone records. The limited food and clothing shops sell quality goods at inflated prices. The East Berlin centre, with its easy access from the Alexanderplatz S-Bahn station is nevertheless a major attraction for tourists who, apart from queueing to visit the TV tower, spread themselves across the (perhaps excessive) open spaces and use the numerous (if still insufficient) catering facilities. A marked contrast with the West Berlin centre is the planned attempt to bring residential development not merely to the immediate fringes of the central area, but over it on the higher storeys.

South of the restored Unter den Linden, the former business and administrative quarter of the Friedrichstadt has until recently been neglected. Surviving government buildings on the Wilhelmstrasse (Otto-Grotewohl-Strasse) and some surviving bank buildings on the Behrenstrasse were reused, but otherwise this quarter, rather remote from the remainder of East Berlin and cut off on two sides by the Berlin Wall, retained many ruins and vacant lots (not entirely unlike the 'Wall fringe' of West Berlin a few hundred metres away). After all, even the considerable GDR bureaucracy and security machine scarcely need for their own 17 million people a government quarter created to meet the needs of a population of 70 million. Until the 1980s only the Leipziger Strasse received a historically insensitive redevelopment as a street of towering residential blocks (in part reserved for foreign diplomats), its former central retailing functions abandoned. On the other hand, the historic buildings of the Gendarmenmarkt (Platz der Akademie) have been under careful restoration for years; the former National Theatre has already been reopened as a concert hall, and work on the French Church is substantially complete. What is new in this quarter is the beginning of an extremely sensitive rebuilding, using pre-cast concrete units as is general in the GDR, but respecting the scale and style of the pre-existing buildings (the contrast with the giant buildings along the Friedrichstrasse is very marked). Close inspection is often needed to decide whether a building is an original structure in stone, or a modern replacement; the surroundings of the Platz der Akademie have received particularly sympathetic attention.

The western side of the Friedrichstadt has always contained a number of foreign embassies, in addition to ministry buildings, and this is still the case. In addition to the vast 1950s Soviet Embassy on the Unter den Linden, the modern architecture and scale of the embassies of Czechoslovakia and North Vietnam put to shame the modest presence of the British Embassy, which does however have the advantage of a shop-window on the Unter den Linden. (Its former site has hitherto lain vacant near the Brandenburg Gate, but may become involved in GDR plans for redevelopment along the Berlin Wall.)

The north–south Friedrichstrasse axis is marked by attempts to return it to its former function in relation to the tourist trade; two major hotels and associated restaurants and cafés already mark the Linden crossing. Whether it will ever become, as Erich Honecker is reputed to believe, one of the major shopping streets of Europe, remains to be seen, but at least the utilitarian buildings of the 'Checkpoint Charlie' crossing point at its southern end have been given a facelift.

7.9 Urban renewal

Very belatedly, attention is also turning to the improvement of the pre-1914 residential fabric of East Berlin. Hitherto the ring of older buildings surrounding the Berlin urban core has been largely neglected; it has been much more economical to use prefabrication to build vast new peripheral settlements than to find craftsmen to repair existing buildings. Matters are made worse by the fact that many apartment houses are still in private ownership; given the extremely low level of controlled rents it is impossible for their often aged owners to finance repairs. In any event, it is extremely difficult for private individuals to obtain building materials or the services of craftsmen.

Now attention is turning to the difficult process of restoration and modernization; gradually the Wilhelmian ring is being given the same sort of treatment as in West Berlin, but the process is extremely lengthy and bureaucratic. Given the near-impossibility of private repairs, whole streets of houses stand deteriorating, with trees growing out of their gutters and the plaster falling off their façades, until the blessed moment comes when the repair brigade arrives to deal with a whole block at one time. Then buildings are ripped out from around interior courts to make gardens and play areas, apartments are provided with modern kitchens and bathrooms, heating and hot water laid on, and façades replastered and painted in the original colours.

An area that has needed particular attention is the old inner fringe, between the former wall of the inner city and the former Customs Wall. Building here was extremely heterogeneous, with much intermixture of industry, and many vacant bombed sites. In spite of its location immediately to the north of East Berlin centre, this belt has been neglected for forty years. Only with the approach of the celebrations for the 750th anniversary of the city was a start made on upgrading the building quality of this run-down area.

7.10 Secondary centres in East Berlin

The belated rebuilding of the East Berlin centre meant that for an even longer period than in West Berlin consumers relied on secondary centres. To a much greater extent than in West Berlin they still do, because of the primarily representational function of the East Berlin centre. Even the large department store on the Alexanderplatz, like its twin at the East (Main) Station, tends to be scorned by 'real Berliners' as being too full of provincial visitors. Another reason is deliberate planning policy, which involves (as in other cities in countries of Soviet-style socialist type) the provision of an appropriate level of retail facilities as close as possible to the population served, not concentrated in a single central area.

This means that in East Berlin there are in theory three levels of secondary centre (Heineberg 1979a). At the base, there is the residential-neighbourhood centre (*Wohnkomplexzentrum*), ideally serving a population of 8,000 to 10,000 (exceptionally 25,000). This should above all contain a supermarket (*Kaufhalle*) selling foodstuffs in daily demand, plus some basic consumers' goods like light bulbs and crockery. Such a centre would also contain a polyclinic for outpatients and a pharmacy, infant and secondary schools, a restaurant, a branch library, a post office, hairdressers, and a reception point for simple services, such as laundry, dry cleaning, shoe repairs, and the repair of electrical appliances. The experience of the present writer of residing first in the Leninallee and later in Marzahn would confirm this general picture of neighbourhood services so far as new settlements are concerned, except that the polyclinic would seem to be related to a higher level in the hierarchy of centres.

The next level, of residential-district centres (*Wohnbezirkzentren*), serving between 20,000 and 60,000 inhabitants, adds to the services mentioned above specialist shops, for example selling dairy produce. Finally city-district centres (*Stadtbezirkzentren*) should have specialist shops selling goods in periodic demand and a department store with a full range of goods capable of satisfying the demands of a market area with more than 35,000 inhabitants. One such centre was belatedly created to serve the Leninallee–Ho-Chi-Minh-Strasse *Grosswohnsiedlung*. As well as stores it includes sports facilities, a swimming-pool, and a restaurant complex, all in a landscaped lakeside setting.

This theoretical hierarchy is really only in place in the new settlement areas, and then only incompletely. East Berlin inhabitants still rely to a large extent on the pattern of secondary centres that emerged from the Second World War. In part these are based on radials through the Wilhelmian residential ring (Schönhauser Allee, Karl-Marx-Allee), in part on nuclei absorbed into the urban expansion of the city (Pankow, Weissensee, Köpenick). There is none of the retailing boom associated with, for example, the Schlossstrasse in West Berlin Steglitz, but the older secondary centres have a high (if diminishing) number of private shops, and are generally esteemed for providing a range of specialist retail outlets.

7.11 The future spatial development of East Berlin

The post-war development of East Berlin has fitted into a generally radial-concentric pattern, which is regarded as spatially economic; indeed, the locating of the new *Grosswohnsiedlungen* into the hitherto empty eastern sector, between the northern and south-eastern axes of urban advance, has made the urban structure even more compact.

Although the boundary between East Berlin and the remaining territory of the GDR is far less meaningful than the boundary of West Berlin, nevertheless the new *Grosswohnsiedlungen* have hitherto been founded exclusively on East Berlin territory. It is probable that the Berlin *Magistrat* is not averse to keeping all population increase within its own territory, thus increasing its influence among the GDR *Bezirke*. The alternative of boundary extension would certainly invoke protests from the western Allies at the modification of an area over which they hold that they have special rights. It is a change that the GDR government would make if unavoidable, but would not undertake lightly.

Plans for the further development of East Berlin region, going beyond its present boundaries, involve a three-pronged advance to the south-east, east, and north, separated by green wedges, with major recreational areas to the north and south-east.

7.12 Urban development in the agglomeration field

In 1983, between 60,000 and 70,000 persons who worked in Berlin lived outside its boundaries (Zimm and Bräuniger 1984). About 86 per cent of them came from the *Kreise* bordering the city on its eastern side, which now make up the reduced crescent-shaped commuter field of present-day East Berlin (fig. 7.5). In 1975 there were about 49,300 commuters from a zone with a journey-time to central East Berlin of up to ninety minutes. Of these 64 per cent came from a ring of *Gemeinden* immediately bordering East Berlin, having a journey time of up to forty-five minutes and typically sending at least 30 per cent of their working population to the capital. A further 27 per cent came from *Gemeinden* with a journey time of between forty-five and sixty

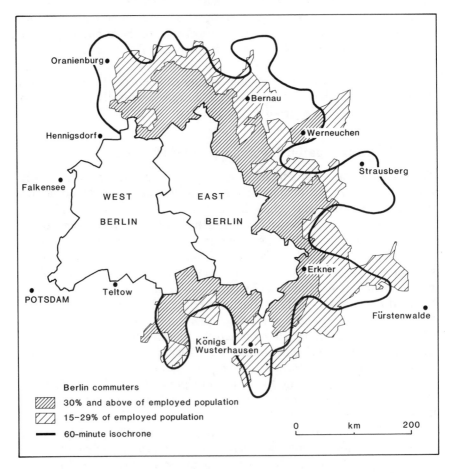

Figure 7.5 The East Berlin commuter field (Source: Spitzer, H. 1985)

minutes, in which commuters typically made up 15–29 per cent of the working population (Kalisch 1980; Spitzer 1985). This sixty-minute zone takes up a pronounced star shape, with arms extending along the radiating rail axes.

The two zones taken together are relatively urbanized; most *Gemeinden* have at least 2,000 inhabitants, there are various small towns of up to 5,000 inhabitants, and four *Kreis* capitals (Oranienburg, Bernau, Strausberg, and Königs Wusterhausen). There are considerable stretches of pre-war low-density, single-family housing, as well as areas of summer-houses (dachas), some of which in times of housing shortage have been transformed into permanent residences. Additional housing has mainly been built in relation to existing or new industrial developments. Some of this is also of low-rise, owner-occupied type, but mostly the usual medium- to high-rise prefabricated apartment blocks are constructed.

The *Kreis* town of Bernau provides an interesting example of the apparently contradictory nature of GDR policies on urban conservation. The small *Altstadt* of Bernau is surrounded by an extremely well preserved defensive system composed of multiple banks and ditches, which is destined to be carefully restored in the near future. In contrast with the extremely careful (and extremely costly) restoration of certain old towns of the GDR, the *Sanierung* of Bernau has taken the form of the virtually total clearance of the old town within the defences. The half-timbered houses have been replaced by apartments built of the customary prefabricated units, but limited to not more than three storeys in height, and with an obvious attempt to provide an acceptable appearance and a sympathetic layout. Among the few surviving buildings are the church and an attractive old inn.

Those parts of the pre-war urban field of Berlin that now lie in the 'shadow' of West Berlin have undoubtedly suffered. North of West Berlin the sixty-minute isochrone just reaches the northern Havel valley, where Hennigsdorf, Velten, and Oranienburg in any event offer important employment opportunities. To the south it falls short of Teltow. *Kreis* Nauen in particular has suffered persistent population loss since the early post-war years, a fate shared to a greater or lesser degree by other areas fringing West Berlin.

The interesting exception to the generally somewhat gloomy picture is provided by Potsdam. This originally Slav settlement, surrounded by rivers and lakes, became a German castle and market town in about 1160, and then sprang into prominence after 1660 as a favoured residence of the rulers of Brandenburg-Prussia. It was also a garrison town, closely associated with the development of the Prussian army. Although the baroque *Stadtschloss* (town palace), was demolished following wartime damage, there is still a remarkable group of rococo and classical palaces, with related monuments, notably the group of the Neues Palais, the Orangerie, and Schloss Sanssouci in the park of that name. Under the influence of the ruling house, Old Potsdam received remarkable planned extensions in the baroque period, a large part of which has survived. There is also a fine collection of public buildings and villas dating from the nineteenth and early twentieth centuries. It is not surprising that Potsdam in its attractive setting of lakes and forests is an important tourist attraction.

Potsdam still retains its military links and has some significant institutions of higher education and science. These include the Educational University at the Neues Palais, the 'Akademie für Staats- und Rechtswissenschaft' in Potsdam-Babelsberg (a kind of GDR 'Ecole nationale d'administration') and the extensive buildings of the Observatory (now Central Institute for Astrophysics of the Academy of Sciences), including Erich Mendelssohn's 1920–1 Einstein Tower, a rare example of Expressionist architecture. Potsdam's rise in population over the last fifteen years can, however, be linked to its function as capital of one of the GDR's *Bezirke*. In this respect it has benefited from being on the 'wrong' side of West Berlin, being able to grow in independence as an administrative and commercial centre. The pedestrianized Klement-Gottwald-Strasse in the baroque town provides what is by GDR standards a lively shopping precinct visited by people from all the surrounding countryside.

The Berlin population | 8

8.1 The population development of Greater Berlin

Berlin reached its maximum population of 4,489,700 in the war year 1943. While the needs of administering a vastly expanded territory and of the war effort had continued to attract additional population up to this point, already forces were at work which were to contribute to subsequent decline. Although the negative natural variation of the population that had appeared after 1925 was reversed by National Socialist population policy, negative figures returned in 1942 and have remained so far as West Berlin is concerned ever since. Negative also were the figures for people of Jewish descent, who disappeared either through emigration or to the death camps; as compared with the 160,600 acknowledged Jews of 1933 (there were very many more of Jewish or part-Jewish descent), the Jewish community in West Berlin numbered only some 5,000 members in the 1970s, and in East Berlin only about 500. There was a parallel haemorrhage of political 'undesirables'. An attempt was also made from the end of the 1930s to check Berlin's population growth by means of restrictions on migration to the city (fig. 8.1).

From 1943 to 1945 Berlin experienced a catastrophic population decline to 2,807,000. Causes included the enlistment of men and women for military service, the call-up of men and women for labour service elsewhere in German-held territory, the official evacuation of children and old people, the privately arranged movement of other members of the population, and the evacuation of industrial establishments and official offices as a precaution against air raids. There were also direct losses of population from bombing and during the final assault of the Red Army upon the city, and above-average mortality through typhus and other diseases at the end of the war. There was then at first a fairly rapid recovery of population with the return of evacuees and the reasonably rapid release of prisoners of war by the

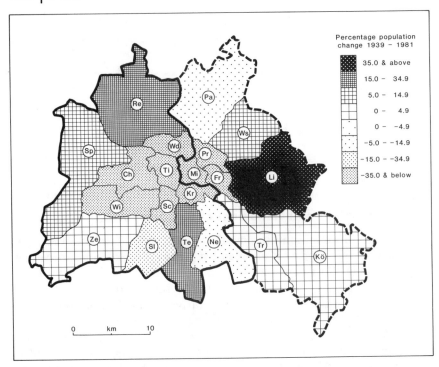

Figure 8.1 Population change 1939–81. (For names of districts (*Bezirke*) see fig. 2.1.) (Source: Müller 1985)

western Allies. By October 1946 total population had recovered to 3,187,470, all but 6 per cent of whom claimed to have been resident in 1939. In the immediate post-war years the shattered city, with its high level of unemployment and restrictions on entry was not an attractive destination for the floods of Germans expelled from territory further east, who went initially to the rural areas of the Federal Republic. Of the relatively small numbers of immigrants who came to Berlin, the largest contingent came from the then Soviet zone of occupation.

After the completion of the partial post-war recovery, population drifted slowly down, from 3,336,500 in 1950 to 3,051,000 in 1981 (Müller 1985). Although the complete isolation of West Berlin came only in 1961, West and East Berlin were attached from 1949 to two greatly contrasting political, social, and economic systems, and the development of their populations needs to be studied separately.

8.2 Population development in West Berlin

The period between the imposition of the blockade in June 1948

Table 8.1 *Population of the Berlin agglomeration 1939–85*

	1939	1950	1971	1981	1985
West Berlin	2,750,494	2,146,952	2,083,987	1,888,669	1,860,100
East Berlin	1,588,262	1,189,074	1,087,982	1,165,677	1,215,300
Greater Berlin	4,338,756	3,336,026	3,171,929	3,054,346	3,072,500
Agglomeration field	358,700	331,400	560,400	561,400	–
Agglomeration	5,188,100	4,200,000	4,035,100	3,925,000	–

Source: Statistisches Jahrbuch Berlin; Müller (1985).

and the year 1952 was one of limited mobility; West Berlin's population declined again by about 15,000 during the blockade period, reflecting the difficulties of food supply. At the moment when the division of Germany was formalized by the creation of the Federal Republic and the GDR in 1949, West Berlin (1950 figures) had somewhat over 2 million inhabitants. Unemployment and scarcity of housing continued to make it unattractive as a migration goal, although there was some continuing drift of people from the GDR.

Then in 1952 the GDR began the process of sealing its boundaries both with the Federal Republic and with West Berlin. For the next nine years, however, the peculiar occupation status of Berlin was respected, and movement between the two parts of the city continued more or less unhindered. Berlin thus became the main channel for population migration out of the GDR. For many, no doubt, the reason for migration was political, but this was also a time when the 'West German economic miracle' was gathering momentum, and the migrants, many of whom were in the economically active age group and many of whom had skills to offer, knew that they could hope for a higher standard of living in the west. About 1.5 million people moved from the GDR into West Berlin, mostly using the S-Bahn, which still provided a unified service linking both parts of the city; only about 109,000 moved in the opposite direction. Residential restrictions in West Berlin were still in force, so that the majority of immigrants were given only temporary shelter in transit camps before being flown out for resettlement in the Federal Republic. Those who stayed in West Berlin were to a considerable extent 'old Berliners', from East Berlin or from the city's rural fringe, who had a sentimental attachment to the city and could obtain shelter with friends or relations and so avoid being sent on to the Federal Republic. In the process, West Berlin in 1953 reached its highest post-war population figure, of approximately 2,386,000, but the long-term decline began again in 1958.

A new phase was entered in 1961, when West Berlin was completely sealed off and movement from the GDR was reduced to a trickle of old-age pensioners rejoining their families and a mere handful of escapers. So far as population developments are concerned, the GDR drops out of the West Berlin story at this point. An immediate consequence was that in the period 1962–70 net population gain through migration was at 72,000 about a third below the figure for 1950–61. There was also a change in the sources of migrants. A greatly increased figure of 173,000 immigrants, 22 per cent of the total, came from foreign countries (see section 8.3). Ten times as many immigrants as in the previous period (537,200) came from the Federal Republic. The amplitude of movements to and from Berlin must, however, not be overstressed; they were far less than those experienced by the great cities partaking at the same time in the economic regeneration of the Federal Republic. The census gives a migration gain of 58,643 persons for the period 1961–70, but total population declined because of a negative figure for natural variation (see section 8.3.1).

In the period 1970–84 the generally downward trend in total population continued. Throughout the period net change by migration was only intermittently positive, peaks in the late 1960s to early 1970s and again in the late 1970s to early 1980s coinciding with peaks of immigration from outside the Federal Republic (fig. 8.2). A further migration peak beginning in the mid-1980s appeared, however, to be significantly different in origin. In the years 1985 and 1986 there was a positive migration balance of the German population, through movement to and from the Federal Republic. An encouraging feature was that the immigration of Germans was particularly high for the age-group 18–30. By 1985, also, the migration balance for the foreign-born population, which had been negative for two years, again became positive. These migration gains were sufficient to cancel out the rising surplus of deaths over births in the German population, so that the total population of West Berlin began to rise again. Were these figures to be maintained, they would represent an important turn-around in the population development of the city (Table 8.2).

8.3 The foreign population

8.3.1 Population composition

When in 1961 the construction of the Berlin Wall meant that industrial expansion in the Federal Republic could no longer be fuelled by a continuing flow of migrants from the GDR, employers turned to foreign countries for a supply of labour, mainly unskilled. Berlin had the additional problem that it was deprived overnight of the workers who had commuted daily from East Berlin and from urban-fringe communities in the GDR. The city accordingly joined in the recruitment

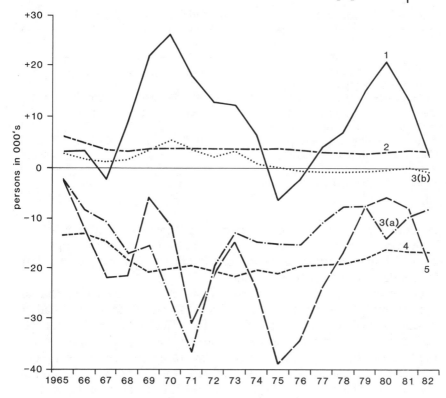

Figure 8.2 Components of population change in West Berlin 1965–82
(Source: Backé 1984)

of foreign workers, although somewhat more slowly than the Federal
Republic, because its economic recovery was slower; until 1968 there
was still only the modest total of 18,754 registered foreign workers.
Demand for labour then increased, following the minor economic
recession of 1966–7, but precisely because of the improving economic
position it was difficult to attract workers from the Federal Republic,
and recourse to foreign workers was the only apparent alternative. A
particular characteristic of the Berlin economy has always been its

Table 8.2 *West Berlin: components of population variation 1965–85*

	1965	1975	1984	1985
Births per thousand	11.8	8.8	9.6	9.7
to foreign population, per thousand total births	286	276	202	437
Deaths per thousand	18.0	19.5	17.5	17.6
Natural variation per thousand	− 6.2	− 10.7	− 7.9	7.9
Migration balance, thousands	+ 10.6	− 17.7	+ 8.7	+ 26.2
Total variation per thousand	+ 4.4	− 28.4	+ 0.8	+ 6.2

Sources: Die Kleine Berlin Statistik, 1985
Statistisches Jahrbuch Berlin

demand for female workers, not only in the tertiary sector but in industrial branches such as the manufacture of electrical equipment, so that employment could be offered not only to male immigrants but to female members of their families.

Because the economic recovery of West Berlin lagged behind that of the Federal Republic, foreign workers attracted to Berlin had a distinctive ethnic composition. By the time Berlin was in the market, the countries which had provided the first waves of foreign workers for the Federal Republic were no longer important suppliers; Italy, Spain, and Greece had no further substantial reserves of labour. Berlin had to look further east and south-east; by far the largest contingent of foreign-born population is derived (1985 figures) from Turkey (45 per cent of the total) and Yugoslavia (Table 8.3). All other groups are small; even the (non-military) British account for 2.4 per cent, and (non-military) citizens of the United States for 2.7 per cent.

Table 8.3 *West Berlin: principal immigrant groups as a percentage of total foreign-born population*

Country of origin	1981	1982	1983	1984	1985
Turkey	49.0	48.0	47.9	45.5	44.9
Yugoslavia	13.5	12.6	12.4	12.5	12.2
Poland	2.8	3.4	3.9	4.6	4.7
Italy	2.9	2.9	3.0	3.0	3.0
Greece	3.0	3.0	2.9	3.0	2.9
Other	28.8	30.1	29.9	31.4	32.3

Source: Statistisches Jahrbuch Berlin.

It is typical of the Berlin foreign population that particular groups experience high rates of immigration at times when their countries of origin are experiencing political or economic difficulties, for example Poles after the difficulties associated with the 'Solidarity' episode. Apart from the Poles, particularly high rates of increase were in the 1980s associated with immigration from the Near East and Asia, such as the Lebanon (1.3 per cent of the foreign-born population), Iran (1.6 per cent), Pakistan, Vietnam, and Sri Lanka. Totals for the countries in this group remain in absolute terms small, but the immigrants concerned are highly 'visible' and in the mid-1980s were swept up into the controversy surrounding the right to political asylum (see section 8.3.2).

According to federal government statistics, West Berlin has the highest proportion of foreign residents of all the federal states, with 12.5 per cent(1982), as compared with a federal average of 7.6 per cent. This high figure partly reflects the fact that West Berlin is a city-state; the foreign population of the Federal Republic is particularly concentrated in the great industrial cities. When measured against these, West Berlin appears less exceptional; although having in absolute terms the highest number of foreign residents of all cities of the Federal Republic, in terms of the proportion of foreign residents it occupies only the ninth position, well behind Frankfurt, Stuttgart, or Munich.

The foreign-born population has contributed significantly to a containment of the long-term fall in the West Berlin population. At the end of 1982, for example, there were 1,724,000 persons of German nationality living in Berlin, a decline of 248,000 (12.6 per cent) on the end-year figure for 1973. At the same time there were 248,000 foreign-born residents, representing a rise of 70,000 over 1973 figures (an increase of 39 per cent). The contribution of foreign immigrants to the demographic structure of the Berlin population was perhaps even more significant (see section 8.5).

8.3.2 The question of political asylum

The Basic Law (constitution) of the Federal Republic guarantees the right of asylum to those forced to leave their countries of origin because of political oppression. The provision was regarded as some sort of return for the shelter that foreign countries had given to 800,000 victims of National Socialist ethnic and political oppression. It also fell into line with a desire to offer succour to the victims of a European struggle against communism. That it should come to be used as a means of entry into the Federal Republic primarily by people from African and, above all, Asian countries of low living standards was certainly not foreseen.

West Berlin was the major point of entry for those claiming political asylum in the Federal Republic; in the first half of 1986, 73 per cent of entrants took this route. Because of the western standpoint that Berlin should be considered to be an undivided city and that the treatment of East Berlin as part of the territory of the GDR is an illegality, there are no customs or immigration controls on the boundary with East Berlin. Anybody managing to obtain access to East Berlin accordingly has the possibility of unrestricted entry into West Berlin. It was an open secret that the movement was frequently organized by unscrupulous promoters, who in return for all that their clients possessed or could borrow would guarantee to deposit them on the streets of West Berlin, which were no doubt depicted in the customary fashion as being paved with gold. The usual route was by means of the Soviet airline Aeroflot or the GDR airline Interflug to the Berlin-Schönefeld Airport, where the transients would be allowed to pass the normally meticulous GDR frontier control whether they had a Federal German entry visa, an obviously forged visa or no visa at all. The motives of the GDR authorities in this strange process were entirely opportunistic; they gained a little foreign exchange from the air flights, but more significantly hoped for the political advantage of provoking West Berlin into instituting frontier controls on the crossing points of the Berlin Wall, thus accepting that this was an international boundary not just an internal division of Berlin.

Fortunately, by October 1986 the GDR government had been convinced that there were greater political and economic advantages to be had from being co-operative rather than otherwise, and announced that visas for their ultimate destination would be required of all transit passengers at Schöneberg.

Once in West Berlin, it was possible for migrants to register as applying for political asylum. The federal and West Berlin authorities then had the task of deciding whether an applicant was merely trying to obtain economic betterment or had a genuine case for political asylum, a difficult enough task when most applicants, such as Iranians or Tamils from Sri Lanka, could plausibly fall under both categories. It could quite normally be two years or, with use of the various possibilities for appeal, up to six years before there was any question of expulsion to the country of origin.

The numbers applying for political asylum in West Berlin rose steadily in the 1980s, from 15,173 in 1978 to 23,000 in 1985. By 1986, when 15,034 arrived in the first half-year and hundreds might arrive on a single day, the situation was considered to be at crisis point. The largest group (1985) consisted of Tamils from Sri Lanka, followed by Iran (3,520, mostly young men trying to avoid service in the Ayatollah's army), Lebanon (3,209), Ghana (2,037) and India (1,780), together with 1,717 stateless Palestinians. These arrivals, coming at a time when there was no demand for additional labour, caused a great deal of adverse popular criticism, and demands from right-wing political circles for a revision of the constitution. It must be emphasized, however, that the figures of arrivals are still extremely small compared with the numbers of Turkish and other foreign workers deliberately imported to undertake menial and unskilled work at a time when the Federal Republic and Berlin were desperately in need of additional labour.

8.3.3 Social and spatial aspects of the foreign population

Although there are something like 160 different nationalities present in West Berlin, it is very much the Turkish population that dominates public attention as a 'problem'. Far more than the groups of immigrants from southern Europe or Poland, the Turkish population is perceived as culturally distant, set apart by language, religion, social customs, family structure, physical appearance, and female costume. So far as possible the natives ignore the Turkish population, who if visible are liable to irritate, as for example when their large family groups take possession of the lawns of the Tiergarten for barbecues on summer Sundays, and the smell of roast lamb, that very non-German dish, fills the air.

The mechanisms of socio-spatial segregation are familiar enough in relation to the immigrant populations of all the great cities of Western Europe. The Turkish population, having been introduced mainly to do the menial and unskilled jobs unwanted by the resident German population, has generally low incomes. The Turkish children often came late into the school system on arrival from their homeland, and so were hampered by lack of knowledge of German. Even those beginning school at the customary age of 6 lack a German-speaking home background, so they tend to leave school with poor qualifications, perpetuating a cycle of deprivation. Some Turkish families, probably now a diminishing number, may also be trying to save money for an eventual return home.

The Turkish population has accordingly looked for low-cost accommodation, for which competition from German families is minimal. In Berlin terms, this means concentration into the dense-packed apartment buildings (*Mietskasernen*) of the Wilhelmian residential ring built in the years prior to 1914 (see section 6.3). The dwellings were substandard, heated only by stoves, without bathrooms and often with only shared WCs. This was very much a Burgess-type 'zone of transition', although the extent of decay was heightened by the presence of the Berlin Wall, cutting the natural links between these quarters and the traditional core of the city (fig. 8.3).

The Turkish residents are found above all in Kreuzberg (20 per cent

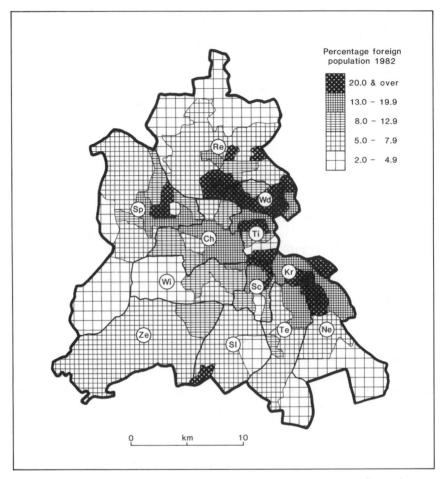

Figure 8.3 West Berlin: foreign population as a percentage of total population. (For names of districts (*Bezirke*) see fig. 2.1.) (Source: *Berliner Statistik* 7/84)

of the total population) and Wedding (14.8 per cent), also in Neukölln, Schöneberg, and Tiergarten. These five districts have three-quarters of the total Turkish population. They are shared with other foreigners (a further 10 per cent of the population of Kreuzberg, for example) and with economically weak elements from the German population, such as the elderly, the unemployed and single-parent families. Smaller concentrations of Turks are found in association with outlying industrial plants, which often have residential homes for foreign workers, as at Spandau. The West Berlin index of segregation for all civilian foreigners (including diplomats and other high-income people) reached a maximum 37 in 1976, but had dropped somewhat to 35 by

1981 (on a scale from 0 = no segregation, 100 = total residence of foreigners in exclusive ghettos). By comparison, the segregation index for academics in West Berlin at the same time was 34! (Backé 1984). The small statistical decline in the segregation index is probably related to the dispersal of population from some residential blocks, notably in Kreuzberg, in the course of urban renewal (*Sanierung*) projects.

Kreuzberg, the smallest of the Berlin *Bezirke* but the most densely populated, isolated well to the east of the city immediately abutting on the Berlin Wall, and abandoned by large sections of the German population, is the largest and best-known Turkish residential area. While, as noted above, for the district as a whole the Turkish population is about 20 per cent, it can make up 50–60 per cent in particular blocks, and individual apartment houses can be totally in Turkish occupation.

Scholz has outlined the build-up of areas of Turkish concentration (Scholz 1985). The initial Turkish 'guest workers', almost entirely male, were accommodated by their employing firms in buildings of all kinds, varying from flimsy huts to purpose-built hostels, which were distributed throughout the city. To overcome their isolation they were accustomed on public holidays to gather in large groups at customary meeting places, of which the best known were the Zoo railway station and Wittenberg-Platz. At such gatherings they could exchange Berlin news, question those newly arrived from the homeland, and buy Turkish newspapers.

In the middle of the 1960s the numbers of immigrants increased: families began to be brought over and long-term rather than temporary accommodation was required. By the beginning of the 1970s the areas of Turkish concentration in the 'transition zone' areas such as Kreuzberg or Wedding were apparent. There was then a rapid build-up, not only with continuing male immigration until the mid-1970s, but with the reconstitution of families. This produced a demographic revolution, and plentiful possibilities for conflict, in areas previously dominated by an aged, and predominantly aged-female, German residual population.

There was also an obvious impact on the schools: in 1981–2, 55 per cent of pupils in the *Grundschulen* (ages approximately 6–11) and 58 per cent in the *Hauptschulen* (ages 12–16) in Kreuzberg were foreign; in individual schools the figure reached 70–80 per cent (Backé 1984). One can imagine the reactions of the parents of the minority of German children to the educational implications of these figures.

It is only necessary to walk down one of the main streets of Kreuzberg to see how the district offers much more than low-rent residential accommodation; because of the concentration of Turkish population a range of support facilities can also be provided. German shopkeepers have given way to Turks, especially for food, where Turkish specialities, the provision of ritually killed meat, and the absence of pigmeat products are obviously important. The small German *Kneipen* (taverns) that were so numerous in these apartment-house districts are now Turkish cafés, where the same men seem to sit the day long in a cloud of tobacco smoke. A study of the Kottbusser Tor quarter of Kreuzberg showed that cafés and restaurants constituted the numerically most important group of Turkish businesses, followed by food shops (Scholz 1985). Characteristic of Turkish districts are shops offering a mixture of services, such as travel arrangements, import–export facilities, insurance, banking, video cassette hire, interpreting, letter-writing, and other services of a specifically Turkish flavour. There are also mosques, their associated religious schools, Turkish lawyers, and Turkish doctors.

A highly individual feature of Turkish commercial life is the street market on the Maybachufer of the Landwehr canal, on the boundary of Kreuzberg and Neukölln. Before the Second World War this has been Berlin's greatest weekly street market, featuring mainly smallholders from Berlin's surrounding countryside, selling vegetables and fruit. By 1961 they could no longer come to West Berlin, and in any event the arrival of the supermarket was changing the nature of retailing; the street market was clearly withering away. It has been completely revived as a Turkish market, a form of individualistic retailing perhaps more acceptable to Turkish purchasers of rural origin than the western-style supermarket. The market is held on Tuesday and Friday afternoons, the latter being by far the more important time, because purchases can be made for the weekend if not for the week ahead, and the occasion used for meeting with friends and acquaintances (casual inspection of Turkish purchasers suggests the frequent presence of the males of the family, more accustomed than their womenfolk to dealing with the outside world, able to speak some German, and perhaps in possession of a car). Fruit, vegetables, and herbs are still the most important commodities, as well as grocery items, basic textiles, and household utensils (Scholz 1985).

There is a good deal of evidence that Turkish businesses are increasingly frequented by German purchasers, who are, for example, attracted by the reasonably priced and often exotic fruit and vegetables from the street market on the Maybachufer. Perhaps the oriental atmosphere of the market also makes shopping there an 'experience' for the German visitor. Less exotic but of practical use are the Turkish-run refreshment kiosks to be found on street corners throughout the city.

The most curious example of a deliberate, if not entirely successful, attempt to appeal to non-Turkish customers is, however, the so-called Turkish Bazaar (*Türkischer Basar*). This has been created in the abandoned Bülowstrasse U-Bahn station (the 'underground' railway at this point was built as an elevated section above the highway). The original idea was that the Turkish Bazaar would attract Germans and foreign tourists to buy Turkish craft goods in an oriental atmosphere. The attempt at tourist appeal was emphasized by the way in which it was linked by an antique tram running along the elevated railway track to the next abandoned station, at Nollendorf-Platz, transformed in similar fashion into the Berlin Flea Market (*Flohmarkt*). The Turkish Bazaar did not achieve its aim of appealing to German residents of Berlin and tourists, partly because of its eccentric location in the 'zone of transition' near the Berlin Wall, partly because it was located actually astride the Potsdamer Strasse, at this point constituting Berlin's 'red light district' and centre of drug trading, and partly because stall rents, and hence prices, were high. The project became, in effect, more 'Turkish', appealing above all to Turkish males, who patronized the increasing number of stalls renting video cassettes and its 'Casino', devoted to entertainment functions.

The foreign workers were introduced into West Berlin for economic reasons; only belatedly did the West Berlin senate try to come to terms with the social consequences of the movement. Initial reactions tended to be negative, such as the 1975 attempt to place a ban on additional immigrants settling in the most frequented quarters of Kreuzberg, Wedding, and Tiergarten. The senate also gives financial support to the older immigrants wishing to take their retirement in their homeland. In a positive sense the work of the senate has particularly concentrated on the fostering of the cultural traditions of the immigrant groups, which alone, it is believed, can provide a firm basis for outreach into the wider Berlin community. There is an emphasis on work with Turkish women and girls, who through cultural tradition are apt to be cut off from the outside world. There is a particular concern with the second and third immigrant generations, who are clearly not going to return to the countries from which they are derived, and who need to be helped towards an equal start in adult life. For example, the number of foreign pupils obtaining the regular school-leaving certificate rose from 30 per cent in 1978–9 to 60 per cent in 1982–3. At the same time the number of foreign youths securing apprenticeships tripled.

8.4 The structure of the Berlin population

The way in which West Berlin has, to a significant degree, remained apart from the general development of the Federal Republic is reflected in the structure of its population of German nationality (see fig. 8.3). Clearly, this is a city where the ageing residents correspond more to its previous imperial role than to its present more restricted functions. At the end of 1983, just under 20 per cent of the population was aged 65 or more (Table 8.4).

Table 8.4 *Changing age structure of the Berlin population*

Age	West Berlin (%)				East Berlin (%)				
	1960	*1970*	*1980*	*1983*	*1960*	*1970*	*1980*	*1983*	*1984*
14 and below	13	15	15	14	16	22	20	19	19
15–64	69	64	64	66	67	62	66	68	69
65 and above	18	21	21	20	17	16	14	13	12

Sources: Müller (1985); *Statistisches Jahrbuch Berlin*

If women alone are taken, the figures are even more extreme; women of 65 and above made up 26.5 per cent of all women, and 14 per cent of the total population. The age of the population is reflected in the statistics of natural variation; in 1984, for example, there were 14,612 more deaths than births, but the deaths were related primarily to the German age groups over 65, whereas the births were predominantly to be attributed to the foreign population. What is more, the deaths were overwhelmingly concentrated in the female population which, as noted, vastly exceeded the male in age groups of 65 and above. This marked female excess only partially reflects the normal tendency for women to live longer than men; these are also the women whose partners failed to return from the war. The negative female natural variation (1984) of 10,649 was much greater than the positive balance of female migration (3,038), whereas for males the position was reversed: the negative natural variation of 3,963 was distinctly less than the male net gain through migration of 5,657.

Berlin's population structure has been modified by the process of migration (figs 8.4 and 8.5). At first, most of the immigrant foreign workers were males living alone in hostels. Even in the early 1980s the age/sex structure of the male foreign population was dominated by men in the age range of 25–45, who by this stage had, however, been joined by a corresponding contingent of women in the age range of 20–35 (see fig. 8.5). In the 1970s these immigrant adults had been supported in the population pyramid only by extremely young children. Nine years later there had clearly been no marked continuation of the initial wave of male immigration, reflecting a deterioration in the economic situation and more restrictive immigration policies. The adult bulge had merely moved up the population pyramid, which had meanwhile acquired greater solidity at the base from the birth of children to Turkish and other immigrants.

Under the combined impact of the heavy death rates of the older population and of the arrival of foreign immigrants and their children,

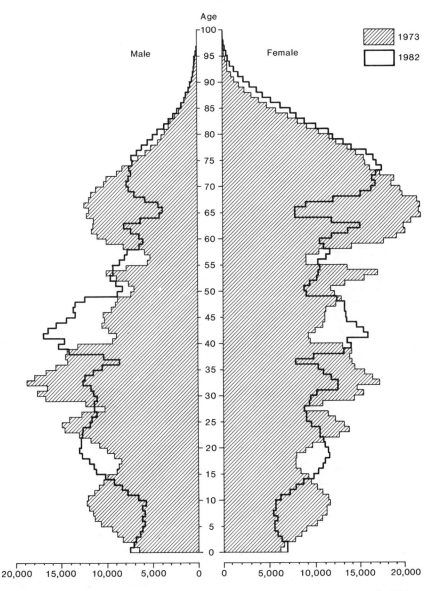

Figure 8.4 West Berlin: German population by age and sex, 1973 and 1982 (Source: *Berliner Statistik* 7/84)

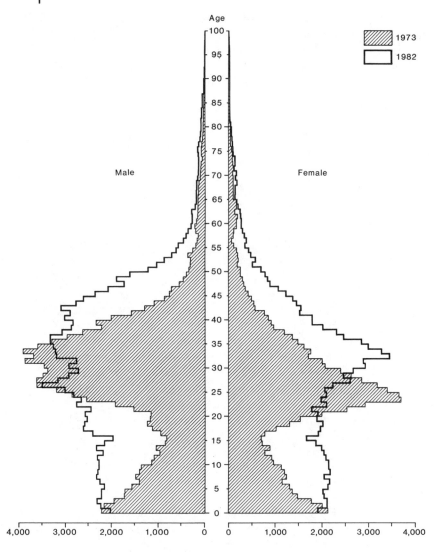

Figure 8.5 West Berlin: foreign population by age and sex, 1973 and 1982 (Source: *Berliner Statistik* 7/84)

the Berlin population structure has in recent years become statistically more normal. The modest birth rate of nine or ten per thousand depends heavily on the foreign population, which contributes 20–40 per cent of the births while making up 12.5 per cent of total population. However, while the crude birth rate of the German population remains fairly static at about 8.3 per thousand, the crude birth rate of the foreign population is tending to drift downwards (15.3 per thousand in

1983, 14.7 in 1984, the lowest figure since the recording of this figure began in 1960). In spite of the over-aged population, Berlin crude birth rates are superior to those of a number of large cities of the Federal Republic (Hamburg, Bremen, and Hanover at 7.7 per thousand). The comparison is not, however, an entirely fair one since the total West Berlin population is necessarily contained within the city limits, whereas with normal city populations it is precisely at the stage when they are having children that families are likely to move out to suburban and peri-urban locations. Berlin birth rates are not, however, greatly out of line even with those of the major states of the Federal Republic, being exceeded only by Baden-Württemberg (10.2), Bavaria (10.1), and Rheinland-Pfalz (9.8).

Although the total population of West Berlin drifted slowly downwards in the twenty years to 1985, the movement was not repeated in the number of distinct households, which did not decline proportionately. In particular, West Berlin has an extraordinarily high proportion of one-person households (Table 8.5). Obviously, this high figure related to the large number of old people living alone, but the changing family structure of our time is also significant. Berlin's 100,000 students certainly make their contribution to this figure. This tendency for a given number of inhabitants to produce an ever-increasing number of separate households is obviously of great significance in the planning of housing.

The structure of population has clearly a close relationship with the size of the labour force, which in the course of the 1960s declined from about a million to 930,000, at a time when the workforce in the Federal

Table 8.5 *West Berlin: age and household structure 1950–82*

Year	Total population (000s)	Population aged 65 and above (%)	Single-person households (%)
1950	2,147	12.5 (9.4)[a]	35.1 (19.4)
1957	2,229	16.3	35.4
1961	2,189	18.2 (11.1)	37.7 (20.6)
1970	2,115	21.4 (13.2)	46.8 (25.1)
1980	1,896	22.1	52.4[b]
1982	1,873	20.6 (14.6)[c]	52.3 (31.3)

Sources: Statistisches Jahrbuch Berlin; Statistisches Jahrbuch der Bundesrepublik Deutschland.
Notes:
(a) Figures in brackets are averages for the Federal Republic, including Berlin.
(b) May 1981.
(c) May 1983.

Republic remained approximately constant. In the 1970s, afflicted by economic decline in the Federal Republic as a whole, the city lost about 100,000 jobs and immigration slackened; the ageing of the population led to a continuing decline in the potential labour force. At this time, only 15 per cent of the population was under the age of 15 (as compared with 23 per cent for the Federal Republic) and 22 per cent were 65 or over, as compared with 13 per cent for the Federal Republic. By the beginning of the 1980s, West Berlin had a workforce of only 835,000 (as opposed to a million two decades earlier). But by the mid-1980s matters appeared to be improving somewhat; unemployment remained high by German standards at 80,000–90,000, but employment, including industrial employment, was growing slightly under the influence of a modest economic recovery, while under the influence of an improved demographic structure and continuing immigration from the Federal Republic the available labour force was also expanding.

8.5 Demographic and social structure of the West Berlin districts

The West Berlin districts are relatively large, ranging in area from 1,037 hectares (Kreuzberg) to 8,936 hectares (Reinickendorf), in population from 276,000 (Neukölln) to 72,000 (Tiergarten), and so can be used in only a very general sense as a basis for social-spatial description. They tend not to fit entirely happily into the actual morphological structure of the city. Nevertheless, it is possible to perceive a degree of differential response to a number of broad changes affecting the city as a whole (H. Müller 1985). They are all linked, for example, by the centrifugal movement of population that has operated in West Berlin as in all western cities, in spite of the city's narrow bounds and lack of access to its natural hinterland (see fig. 8.1). In demographic and social terms the West Berlin districts fall into four broad categories (fig. 8.6).

8.5.1 'Zone of transition'

This type is represented by the districts of Wedding, Tiergarten, Schöneberg, and Kreuzberg, approximately corresponding to the Wilhelmian residential ring in all but its higher-income western sector (see section 6.3). These are areas which, since the division of the city, have been greatly disadvantaged by having their pre-war ease of access to the traditional urban core replaced by distinct isolation from the new West Berlin central business district in the neighbourhood of the Kaiser Wilhelm Memorial Church. The areas tend to be rundown,

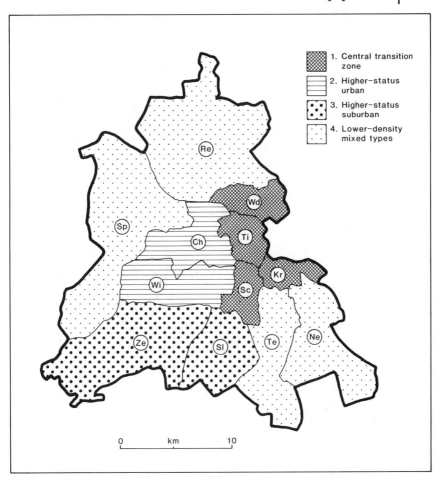

Figure 8.6 West Berlin: social characteristics of districts. (For names of districts (*Bezirke*) see fig. 2.1.) (Source: Müller, H. 1985, Senator für Stadtentwicklung und Umweltschutz, *Flächennutzungsplan FNP 84, Stand 86*, and other sources)

showing signs of changing industrial, commercial, and residential functions.

A significant proportion of the housing stock still consists of high-density apartment houses built in the years prior to 1914, containing generally sub-standard dwellings. There is a general shortage of open spaces, children's playgrounds, and other forms of social provision, especially where these are space-demanding. The German population is heavily dominated by the elderly, but statistically this is masked by the immigrant population, with its high proportions of adults in the economically active years and of children (in Kreuzberg 18 per cent of

Table 8.6 Population change in the West Berlin districts, 1971–85

Urban district	% of city area	1971		1980		1985	
		Population	% of city	Population	% of city	Population	% of city
Tiergarten	2.8	94,600	4.5	73,800	3.9	72,000	3.9
Wedding	3.2	174,300	8.4	135,800	7.6	136,700	7.3
Kreuzberg	2.2	155,800	7.5	130,100	6.9	127,400	6.9
Schöneberg	2.6	163,100	7.8	142,100	7.5	134,600	7.2
Zone of transition	10.7	587,800	28.2	481,800	25.4	470,700	25.3
Charlottenburg	6.3	194,000	9.3	153,600	8.1	145,600	7.8
Wilmersdorf	7.2	147,400	7.1	131,700	6.9	130,700	7.0
Higher-status urban	13.5	341,400	16.4	285,300	15.0	276,300	14.9
Steglitz	6.7	184,700	8.9	169,100	8.9	167,600	9.0
Zehlendorf	14.7	89,700	4.3	84,000	4.4	85,200	4.6
Higher-status suburban	21.4	274,400	13.2	253,100	13.3	252,800	13.6
Spandau	18.0	194,800	9.3	197,000	10.4	192,900	10.4
Tempelhof	8.5	162,300	7.8	163,500	8.6	162,100	8.7
Neukölln	9.4	276,800	13.3	280,000	14.8	276,400	14.8
Reinickendorf	18.6	246,400	11.8	234,400	12.4	229,000	12.3
Peripheral districts	54.4	880,300	42.2	875,800	46.2	860,400	46.2
West Berlin	100.0	2,084,000	100.0	1,896,700	100.0	1,860,100	100.0

Source: Statistisches Jahrbuch Berlin

the 1984 population consisted of children under 16, as compared with the West Berlin average of 13.7). Other socially disadvantaged groups include single-parent families and the unemployed; obviously, these categories of disadvantage frequently overlap, leaving plenty of problems for the city authorities. Occupations tend to be in manufacturing industry or low-grade service employment. Population has fallen dramatically below pre-war figures and continues to show a slight decline, in spite of foreign immigration. Efforts to improve the environment by urban renewal schemes and the creation of small open spaces tend to reinforce population decline.

8.5.2 Higher-status urban districts

This type is represented by the districts of Charlottenburg and Wilmersdorf. At least in their inner sections, they date from the same period of building as the first category, but form part of a distinctive western sector of higher-income development. The area is closely built up, to a considerable extent with apartment houses, but dwellings are adequately equipped. Residential densities are below those of the first category, and gross densities well below, partly because of a better (but still not completely adequate) provision of public open space. A distinctive accent is given to Charlottenburg by its possession of the greater part of West Berlin's central business district, in the neighbourhood of the Kaiser Wilhelm Memorial Church. There are fewer industrial workers, more people with further education, more academics. Households tend to be small, the proportion of elderly persons in the populations still high (22–3 per cent aged 65 or over), and the proportion of the population aged 0–15 below the Berlin average. The area has lost population heavily as compared with pre-war figures; in Charlottenburg population continues to drift downwards, in Wilmersdorf it is relatively stable.

8.5.3 Higher-status suburban districts

Included here are the two districts of Steglitz and Zehlendorf; together with the districts in the following category, they make up the nearest equivalent to a normal suburban ring that is possible in spatially restricted West Berlin. Steglitz and Zehlendorf provide a south-western extension of the sector of higher-status development mentioned above, but are of more recent residential development, having been built up principally in the inter-war and post-war years. Housing is of high quality, with many single-family houses set in their own gardens. Residential density is low, and overall population density extremely low, especially in Zehlendorf, because of the large areas of open land.

The proportion of persons having good educational qualifications is high, the proportion of industrial workers very low, and the proportion of foreigners also very low. Living standards are obviously high. The proportion of elderly persons is just above the Berlin average, of children under 16 near average or a little above (Zehlendorf). The population of Zehlendorf stands at about its pre-war level, of Steglitz somewhat below, but in recent years the populations of both districts have varied little.

8.5.4 Lower-density mixed districts

The remaining peripheral districts of Reinickendorf, Spandau, Tempelhof, and Neukölln are more heterogeneous in structure. Residential densities are moderate, overall densities very low, because of the large areas of surviving open space. Types of settlement include relict villages of agricultural appearance (Lübars in Reinickendorf), an ancient town (Spandau), tracts of mainly inter-war single-family housing of variable quality, 'garden cities' of high residential quality (Frohnau in Reinickendorf), works settlements (Siemensstadt on the borders of Charlottenburg and Spandau), planned low-density housing estates of the inter-war years (Britz in Neukölln, Weisse Stadt in Reinickendorf, Onkel-Toms-Hütte in Zehlendorf), and the high-rise housing projects of the post-war era the largest being Märkisches Viertel (Reinickendorf), Gropiusstadt (Neukölln) and Falkenhagener Feld (Spandau).

The social and demographic characteristics of the inhabitants vary with the age and type of settlement, but the construction in these districts of much of Berlin's post-war housing for economically weaker categories is reflected in the statistics. The proportion of people employed in manufacturing is relatively high, further-educational qualifications rather low, people of 65 and above and children up to 16 are both above average for the city. There are some small pockets of foreign population, related to the large industrial plants found in these outer areas. These are districts which have in general a higher population than in 1939, because of migration from the inner districts, but in recent years they have mirrored the slight population decline characteristic of the city as a whole.

8.6 East Berlin

8.6.1 Population development in East Berlin

West and East Berlin, as they emerged from the chaos of the Second World War, had populations of similar type, marked in particular by a disproportionate number of elderly persons and of women. Since the completion of partition in 1961, East Berlin has not shaken off all traces of these initial characteristics, but its demographic evolution has significantly diverged from that of West Berlin. This can be related to the way in which East Berlin is geographically more normal than isolated West Berlin, remaining politically united with its immediate hinterland in the GDR, to which it has direct, if not totally unimpeded, access. It has also retained the capital functions that have been lost by West Berlin. The fact that East Berlin is governed by a regime devoted to notions of careful and detailed planning with regard to population as in all aspects of life, is also of significance.

Like its neighbour, East Berlin has failed to recover its 1939 population. Post-war recovery reached an initial post-war maximum of 1,209,300 in the blockade year 1949, but declined thereafter, reaching a minimum 1,055,300 at the time of the building of the Berlin Wall. Since that time there has been a slow but steady increase; to this extent East Berlin falls into the family of cities characteristic of the Soviet-style socialist states, where planning tries to concentrate population and settlement. It contrasts with the larger core-cities of Western Europe, which almost without exception have lost population by migration to their rural fringes and beyond.

Unlike the dubious residential attractiveness of West Berlin in relation to the 'mainland' of the Federal Republic, the attractiveness of East Berlin is without rival within the territory of the GDR. As the capital, as headquarters for a whole range of national organizations, as the principal centre for the arts, learning, and sciences, East Berlin offers not only a wide range of employment in the tertiary sector but also, for the ambitious, the glittering prizes afforded by positions of power and influence, far more important in GDR society than money incomes. It is also the greatest single industrial centre of the GDR, and as such has a constant appetite for industrial workers, many of whom begin by living as single persons in hostels, until they have been on the waiting-list a sufficient number of years to qualify for a proper apartment. The government's addiction to prestige building projects (especially marked in the run-up to the 750th anniversary celebrations) has a similar if lesser impact. A further incentive to migrants is that, as capital and showplace for visitors, East Berlin is not only better equipped with cultural and entertainment facilities but better supplied with food and consumer goods than the rest of the country. It is therefore not surprising that East Berlin's share of total employment in the GDR rose by about 9,000 a year in the first half of the 1980s, reaching just under 8 per cent of all GDR employed population in 1985, in which year the Berlin population as a whole had reached 7.3 per cent of the GDR total. The rise was closely reflected in the figures for net migration gain, which in turn has been facilitated by East Berlin's major housing programme (between 1981 and 1985 a total of 79,032 new units and 39,998 modernized units).

Migration was facilitated by the removal in 1974 of the need to obtain special permission to take up residence in Berlin; thereafter annual net migration rates of about 13,000 have been characteristic (Schulz 1986). The migrants were overwhelmingly young adults, 78 per cent of them in the economically active age groups, 66 per cent of them in the age group 18–35 alone (Flauss and Schulz 1983). They were also highly skilled, over a quarter having university-level or technical education. The proportion of males and females in the immigrant populations was approximately equal, in distinct contrast to the existing population. The influx significantly modified East Berlin's demographic structure; the proportion of persons over 65 has been reduced much more rapidly than in West Berlin, and the proportion of persons in the economically active age groups has remained high. Women per thousand males have been reduced from 1,183 (1978) to 1,146 (1983). The arrival of large numbers of young adults has also been largely responsible for a crude birth rate of about 15 per thousand, well above that of West Berlin. The proportion of children up to the age of 15 in the total population was also higher than in West Berlin, while a further consequence was that, by the 1980s, natural increase was positive, more people were being born than were dying. It has to be said, however, that these favourable population developments were obtained at the expense of the remainder of the GDR, which was to some extent drained of active population and potential births.

Table 8.7 *East Berlin: natural variation of population 1965–85*

Per thousand total population	1965	1975	1983	1984	1985
Births	16.5	10.1	15.0	14.2	14.3
Deaths	16.3	14.3	12.7	12.0	12.4
Natural variation	0.2	− 4.2	2.3	2.2	1.9

Sources: Die kleine Berlin Statistik, 1985
Statistisches Jahrbuch der DDR, 1986

What East Berlin does not have is a foreign immigrant population; in spite of what has at least until recently been an acute labour shortage, there was no resort to 'guest workers'. There are small groups of foreigners present as students or industrial trainees, and limited use is made of foreign specialists on specific projects, like the

Polish firms that have done magnificent work on the restoration of historic buildings, but in either case the period of stay is strictly limited, and all concerned return eventually to their own country.

8.6.2 Social and demographic contrasts within East Berlin

In principle, and very largely in practice, there are no social-spatial contrasts in East Berlin. There are, of course, certain exceptions; it is obviously rational to group together those who are deemed to need protection from 'foreign revanchist elements' like high Party members, or from whom the GDR population may need protection, like foreign diplomats, in well-guarded areas of high-class housing in the north of the city. Equally clearly, it is not to be expected that leading musicians, painters, or atomic physicists would be expected to occupy standard-issue apartments in distant Marzahn. That being said, however, the system is highly egalitarian; in principle, all persons needing to be housed go on to the same waiting-list, eventually receiving whatever accommodation is available when they reach the top of the list, whether new, old, or renovated. The number of rooms varies only with family consumption, not with ability to pay, and the location offered is highly random.

If social-spatial variation is thus minimized by the system of housing allocation, the same is not true of demographic variation. Everything depends on the period at which any particular section of the city was constructed. As in West Berlin, the Wilhelmian ring of apartment houses constructed before 1914, as represented by Prenzlauer Berg and Friedrichshain districts, together with Mitte district, containing the traditional inner city, have remained well below their 1939 population (see fig. 8.1). The government's principle that reconstruction of central Berlin should include a high proportion of residential buildings meant that the decline of the relatively small surviving population in the Mitte district was checked in the 1970s, but population decline has continued unabated in Prenzlauer Berg. Friedrichshain, which contained some of the earliest post-war reconstruction projects along the Stalinallee (later Karl-Marx-Allee) and elsewhere, recovered some of its population after the war, but has since the 1960s joined the downward trend. It must be assumed that plans for the considerable renovation of the housing stock in these areas will be accompanied by further population decline. Two other areas of the city offer a pattern of population stagnation or slight decline. The south-eastern districts of Treptow and Köpenick have somewhat more than recovered their 1939 populations, but they have not been selected for major recent housing projects, and in the past decade their populations have drifted downwards. Pankow to the north has not reached its 1939 population;

after some post-war recovery its population has also drifted slowly downwards.

A complete contrast is provided by population evolution in the eastern districts of Lichtenberg and Weissensee, reorganized in 1979 to provide an additional district of Marzahn, and again in 1985 to admit the new districts of Hohenschönhausen and Hellersdorf. It is in these districts that the great high-rise housing projects of the 1970s and 1980s are located (see section 7.7). The result has been that between 1971 and 1985 the population of these eastern districts doubled, from 251,500 to 500,900; a considerable further rise can be expected as projects are completed towards the end of the 1980s. An area that in 1961 has 23 per cent of the population living on 31 per cent of the city's area had in 1985 41 per cent of the population (the area concerned has been slightly increased by boundary changes to 35 per cent of the East Berlin total, but the figures are a fair indication of actual developments). About half the occupants of the new developments came from the crowded inner districts of Prenzlauer Berg and Friedrichshain. The eastern districts also received about 40 per cent of migrants from outside Berlin, the remainder having to contrive niches in the older housing stock of districts such as Prenzlauer Berg until qualifying for something better (Flauss and Schulz 1983).

The people involved in this movement of population into the eastern districts were characteristically young adults, often seeking better accommodation in which to bring up young children. There has been a consequent impact on the population structure. In 1980 the eastern districts had only a marginally higher proportion of people in the economically active age groups than the city as a whole. The great difference lay in a markedly lower proportion of retired inhabitants (13.6 per cent as against 16.5 per cent) and a higher proportion of children (21.1 per cent as against 19.1 per cent). A further increase in the proportion of children was expected in the course of the 1980s, with the completion of further building projects. This 'bulge' of children in the demographic structure has obvious planning implications, particularly with regard to the demands of the educational system.

8.6.3 Population developments in the agglomeration field

Unlike West Berlin, East Berlin is not politically sundered from its natural hinterland. In terms of population development, there is a marked contrast between those parts of the hinterland that have relatively direct access from East Berlin and those that were formerly reached by commuter routes radiating through the now sealed-off enclave of West Berlin. Migration losses have been experienced by the rural *Landkreise* Potsdam and, more particularly, Nauen that lie in

Table 8.8 Population change in the East Berlin districts, 1971–85

Urban district	1971			1980			1985		
	% of city area	Population	% of city	% of city area	Population	% of city	% of city area	Population	% of city
Mitte	2.7	87,952	8.1	2.7	89,975	7.8	2.7	80,300	6.6
Prenzlauer Berg	2.7	199,153	18.3	2.7	189,095	15.8	2.7	166,700	13.8
Friedrichshain	2.4	145,753	13.4	2.4	133,636	11.6	2.4	118,500	9.7
Inner city	7.8	432,858	39.8	7.8	405,706	35.2	7.8	365,500	30.1
Treptow	10.1	132,132	12.2	10.1	122,969	10.7	10.1	111,100	9.1
Köpenick	31.6	128,493	11.8	31.6	125,364	10.8	31.6	119,900	9.9
South East	41.7	260,625	24.0	41.7	248,333	21.5	41.7	231,000	19.0
Pankow (North)	19.5	141,389	13.1	19.5	140,457	12.2	15.2	118,000	9.8
Lichtenberg	19.6	169,830	15.6	6.5	180,221	15.6	6.5	183,200	15.1
Weissensee	11.4	81,672	7.5	9.8	96,023	8.3	7.3	60,500	5.0
Marzahn	–	–	–	14.7	81,789	7.2	14.7	190,200	15.6
Hohenschönhausen	–	–	–	–	–	–	6.8	67,000	5.5
Hellersdorf*	–	–	–	–	–	–	–	–	–
East	31.0	251,502	23.1	31.0	358,033	31.1	35.3	500,900	41.2
East Berlin	100.0	1,086,374	100.0	100.0	1,152,529	100.0	100.0	1,215,300	100.0

Note: * Formed from Marzahn, 1985
Sources: Flauss and Schulz (1983), Magistrat von Berlin, Haupstadt der DDR, 1986

the shadow of West Berlin. By contrast, the town of Potsdam itself, formerly very much a Berlin suburb, reached from the centre by a mere half-hour journey on the S-Bahn, has actually benefited from being accessible from East Berlin only by a tedious journey along a circular railway that seems to be perpetually under repair. It has been able to grow in independence and importance as the capital of Potsdam *Bezirk*, and this is reflected in population growth (111,933 inhabitants in 1971, 132,005 in 1981). The building of extensive new residential complexes has accompanied a marked migration surplus and a positive rate of natural increase (D. O. Müller 1985).

The remaining *Landkreise* bordering Berlin to the north, east, and south, to which access is relatively direct, show a relatively unchanged population when taken together. Calculations by geographers of the Humboldt University relating to the 'agglomeration field' (fig. 7.5) to the north, east, and south of the city give similar results. In the period from 1964 to 1971 there was a modest increase of population, followed by a period of little overall change (population 564,270 in 1983). There were, however, important internal variations (Rumpf *et al.* 1985). The purely rural *Gemeinden* have lost population, primarily by migration to Berlin. Population gain has been experienced, apart from by the *Kreis* capitals, by large *Gemeinden* and small towns (some of them also *Kreis* capitals) on the fringe of Berlin that are centres of industry in their own right and also house Berlin commuters, for example Oranienburg, Bernau, Strausberg, or Königs Wusterhausen. The two *Bezirke* of Potsdam and Frankfurt bordering East Berlin provide 35 per cent of all migrants to the city, 25 per cent from the six *Kreise* immediately bordering the city alone. Well over half of the smaller but growing out-migration from the city is also directed to Potsdam and Frankfurt *Bezirke*.

Because of its extraordinary political division, a total population figure for the Berlin city region is not particularly useful. Compared with other European city regions, west or east, Berlin can only be regarded as an anomaly. At the outbreak of war, in 1939, the city region had a population of 5.2 million, 4.3 million in Greater Berlin. In 1982 the population of the city region had at most 3.9 million, 3.1 million in Greater Berlin, divided between 1.9 million in West Berlin and 1.2 million in East Berlin (Müller 1985). For the record, the population of East Berlin and its 'agglomeration field' was calculated in 1983 to be 1.75 million, of which 1.2 million were in the capital and 0.56 million in the *Umland*.

Life after the Wall | 9

9.1 East Berlin

Of the two sides of the Wall, East Berlin can logically be regarded as having the less problematical future. It enjoys all the prestige of possessing the complex of buildings that made up the pre-1945 capital of Germany. It is today the functioning capital of a state of 16.7 million people, occupying a position of undisputed political, administrative, economic, and cultural leadership. It has direct access to the territory of the state that it governs, even if the sealed-off island of Berlin is an irritating interruption to land communications. The significance of East Berlin as a goal for internal migration underlines the attractiveness of the city both in relation to employment opportunities and to the quality of living that it offers (see section 8.6.1).

9.1.1 Entertainment and recreational facilities

Because East Berlin inherited the heart of the traditional capital, it also inherited the majority of that capital's seats of culture, notably the cluster of great museums on the 'Museum Island', although not the whole of their pre-war contents. The former division of the German-speaking lands into a swarm of principalities left behind not only a great heritage of art treasures in the collections of numerous former rulers but also their widespread availability, rather than concentration in a single national capital. Some collections instantly come to mind, notably those of Dresden, Munich, and (outside the boundaries of present-day Germany) the Habsburg inheritance in Vienna. In Berlin the collections of the Hohenzollerns were further expanded in the years of western political and economic predominance in the late nineteenth and early twentieth centuries. Not only individual objects but entire Babylonian and Greek buildings were transported from areas under Turkish control, where German political influence was strong. When to the richness of the museums is added current activity in fields such as theatre, music, film, literature, graphic art, and architecture, it is clear that no other city in the two German states is so outstanding culturally as Berlin.

Of course, in cultural as in all other matters, West and East Berlin are quite separate. The post-war division of the city's cultural inheritance is part of the general tragedy of Berlin's division (see section 2.7). During the war years the greater part of the museum contents, like the contents of the Prussian State Library, were evacuated to places of presumed safety. A proportion never returned to Berlin, destroyed or looted in the war. Only those surviving items that were discovered in the areas conquered by the armed forces of the Soviet Union eventually came back to East Berlin. The resources are nevertheless still of fantastic richness.

Of all the collections of the Museum Island it is perhaps the Pergamon Museum that draws the crowds. A sacred processional way from the Babylon of Nebuchadnezzar II leads to the magnificent Ishtar Gate, its glowing turquoise bricks set with reliefs of bulls and dragons. Beyond lie the great Greek structures from Asia Minor, including the enormous altar from Pergamum (Pergamon) and the scarcely less imposing Market Gate from Miletus. The building also contains a remarkable Islamic collection. The most imposing of the museum buildings, Schinkel's colonnaded Altes Museum, forms the north-western termination of the Lustgarten; it houses paintings of the twentieth century, drawings and engravings. The National Gallery is devoted to art of the nineteenth and twentieth centuries. The Neues Museum of 1843–7 has stood as a ruin since the Second World War but its reconstruction has been announced; it will probably house the Egyptian antiquities at present in the Bode Museum.

Three contrasting museums out of the many that remain may perhaps be mentioned. On the Unter den Linden the baroque Zeughaus (Armoury) houses the Museum of German History, concentrating in particular on 'the fight of the working-class movement against imperialism, militarism and fascism, until the culmination of the struggle was marked by the creation of the GDR'. South of the Spree, the Märkisches Museum is devoted to the history of the city and its cultural developments. In the east of the city, Schloss Köpenick contains the Kunstgewerbemuseum (Museum of Applied Arts), which in the inter-war period was housed in the Berlin Palace.

The West Berlin monthly entertainments guide *Berlin Programm* for February 1987 listed twenty museums or major exhibition centres in East Berlin; in addition to the permanent collections, sixteen special exhibitions were mentioned. Nineteen smaller galleries were also recorded; no doubt these lists could be extended.

The theatre and music are vigorously supported and massively subsidized in the socialist city. Some of the earliest post-war reconstruction on the Unter den Linden included the Deutsche Staatsoper (Opera House). The often provisional repairs of those difficult years were followed by a thorough restoration in time for the 750th anniversary year. It is indicative of the GDR's changed attitude to German history that the 1955 simple inscription on the portico *Deutsche Staatsoper* was replaced in 1987 by the historically correct inscription *Fredericus Rex Apollini et Musi*. The Komische Oper, in spite of its name, is a second house for 'serious' opera and dance; operetta finds its home in the Metropol-Theater. The classical theatre has its home at the 1848 Deutsches Theater, but there are many more theatres of all sizes and artistic inclinations, including, of course, the Berliner Ensemble 'Am Schiffbauerdamm', on what is now called Berthold-Brecht-Platz. Variety shows are provided in the Friedrichstadtpalast and at the theatre in the Palast der Republik. There are even some satirical cabarets.

Schinkel's Schauspielhaus (National Theater) on the Gendarmenmarkt (Platz der Akademie) was reopened in 1984 after eight years of restoration not as a theatre but with a completely different interior in the style of Schinkel to serve as a concert hall, together with a smaller recital hall and music club. Whatever reservations there may be over the artistic justification of not so much restoring as building anew in a pastiche of a past style, it must be said that the craftsmanship is meticulous, the employment of materials lavish, and the effect imposing. With a maximum of 1,650 seats, the concert hall gives a feeling of warmth and intimacy, and appears to be acoustically excellent. Concerts and performances of popular music are also given in the opera houses and theatres.

In this socialist system, theatre and concert prices are ridiculously low by western standards; opera tickets, for example, range (1987) from just over £1.00 to £5.40. The problem for the ordinary Berliner is to lay hands on them; when the organizations such as trade unions and the tourist agencies paying hard currency have had their pick, there is little left for the individual purchaser.

Because East Berlin includes the historic core of the city, there is no shortage of tourist 'sights'. Although the GDR government has been criticized for one or two actions, such as demolishing the remains of the Berlin Palace to make a parade ground, or blowing up the nineteenth-century brick gasholders, it is in general prolific in expenditure on the rehabilitation of its architectural heritage in Berlin. The modern architecture of the new East Berlin centre around the TV tower is itself a considerable attraction. Public monuments are a particular feature. Not only are figures from the militaristic and imperialist past brought back, like Frederick the Great to the Unter den Linden, but socialist heroes of all kinds are celebrated. Marx and Engels appropriately occupy the heart of the new urban centre. Lenin in red granite towers up 18 m high on the street that bears his name, while the fifty-five tonne, 15 m bust of Ernst Thälmann is the centre piece of the new park and housing project on the former gasworks site that now bears his name. There are monuments for resistance fighters, for participants in the Spanish Civil War, for the fallen of 1848 and November–December 1918, for deceased GDR leaders, and many more. Most monumental of all is the cemetery and memorial of the Soviet dead in the Battle for Berlin, in the Treptow Park.

Recreation is very much a matter of central rather than private provision, and facilities are typically lavish. Some are inherited, like P. Behrens's 1912 boat-house and other sports facilities formerly belonging to the AEG firm on the Spree in Köpenick, or the Weimar-period bathing establishment on the Müggelsee. Others are quite new, such as the vast covered 'Sport und Erholungszentrum' in *Bezirk* Friedrichshain, with ice rink and all manner of other facilities. Young members of the official youth organizations are particularly well catered for, as for example at the 'Pionierpark Ernst Thälmann' in Oberschöneweide, with its 7 km 'Pioneer railway', its varied sports facilities, its club house, open-air theatre, and accommodation for visiting young people. The 'Insel der Jugend' (formerly Abtei-Insel) in the Spree, with its extensive bathing beaches, is another example. For those disinclined to intellectually or physically taxing activities there is the extensive Zoo, opened in 1955, where animals are as far as possible shown in family groups, in natural surroundings.

Unlike the inhabitants of West Berlin, the East Berliners have direct access for informal recreation not only to the beauty of the Müggelsee and other lakes and forests on Berlin territory but to its hinterland beyond. The lucky ones among them even have a summer-house (dacha) somewhere around the city fringe (see section 3.8).

9.1.2 Residential attractiveness

There are other reasons for the residential attractiveness of East Berlin within the context of the GDR. As capital, as headquarters for a whole range of national organizations, as the principal centre for the arts, learning, and the sciences, as the GDR's greatest industrial city, East Berlin offers plentiful opportunities for the talented and ambitious. It also offers a better standard of living. The hierarchical nature of resource provision in the GDR has been referred to by Schöller (1986). Just as the *Bezirk* capitals do better than the smaller towns and villages, so the capital fares best of all. For example, the 1950s

Stalinallee project was declared to be of national interest and thus to be supported by workers from the whole of the GDR. So too in the run-up to the 750th anniversary celebrations the various *Bezirke* in the GDR were required to send building teams to operate in East Berlin. Migrants will be well aware that East Berlin is currently benefiting from a massive housing programme and is better supplied with food and consumer goods than the rest of the country.

9.2 West Berlin

9.2.1 As West Berlin sees itself

It is interesting to discover how, in a leaflet designed to attract new skilled workers to the city, West Berlin sees itself. 'Just a moment', it says, 'do you know that Berlin:

- has so much water, woodland, parks and open space that they make up about 40 per cent of the area of the city?
- has more than 830,000 people employed in approximately 2,300 industrial establishments, 15,000 craft businesses and tens of thousands of service enterprises?
- has about 5,000 restaurants, cafés, and pubs, most of them open all day?
- has about 90,000 students, 7,000 of them from Third World countries?
- has about 5,800 flats provided by the ARWOBAU [Arbeitnehmer-Wohnheimbaugesellschaft mbH] for workers moving to West Berlin?
- since the coming into force of the Transit Agreement can be reached easily, quickly, and in comfort?
- is looking for skilled workers from the Federal Republic in many branches?
- with forty-five theatre stages can offer a varied programme throughout the year?
- has thirteen institutes of continuing education [*Volkshochschulen*] and 100 public libraries, open to all?
- has the longest U-Bahn system of any German city (106 km)?
- enables every employed person to receive a tax-free supplement of 8 per cent of the basic wage?
- fixes the upper income limit for access to social housing higher than in the Federal Republic?
- is a major centre for international conferences and trade fairs, with an all-year season?
- has 1.5 million visitors every year?

- gives financial help to those wishing to set up independent businesses?
- is a major centre of education and science, with 180 research institutes?
- has 1,100 varied retail outlets on the Kurfürstendamm alone?
- helps newly arrived workers in the search for housing?
- gives various financial advantages to specialists migrating from the Federal Republic, as long as no Berlin worker is available for the relevant position?'

9.2.2 Entertainment and recreational facilities

Some aspects of West Berlin's attractiveness will bear further examination. Museums are fundamentally affected by the decision that items of the cultural inheritance derived from the whole of Berlin, West or East, that were sent away to wartime places of safety that came under the control of the western Allies, should be returned to West Berlin and placed under the trusteeship of the *Stiftung Preussischer Kulturbesitz*, which at present controls fifteen museums. Fortunately for West Berlin, it had been decided even before the First World War to supplement the Museum Island by a second museum centre at Dahlem, in the south-west part of the city. The Dahlem complex fortunately survived the Second World War and provided a provisional location for the collections of western art. The long-term purpose of the complex is, however, to provide permanent accommodation for the Museums of Anthropology, Islamic Art, Indian Art, and East-Asian Art.

The second main museum centre is provided by the splendours of the 1695–1790 Schloss Charlottenburg, including the Museum of Prehistory, the Egyptian Museum (with bust of Queen Nefertiti as prize exhibit), and the Museum of Classical Antiquity. A third museum centre is the Kulturforum, far away to the east beside the Berlin Wall. This currently inconvenient location relates to the notion of a *Cityband* and to earlier hopes of a reunification of the city (see section 7.4). Included here are the 1965 National Gallery of Mies van der Rohe, for art of the nineteenth and twentieth centuries, and the Museum of Applied Arts, based on part of the collection displayed before 1939 in the Berlin Palace. Accommodation for the Kupferstichkabinett (Department of Engravings) was under construction in 1987. The idea is to unite all the collections of western art on this site. The Kulturforum includes Hans Scharoun's 1960–3 Philharmonie (concert hall); the Chamber Music Hall was under construction in 1987. The Kulturforum is cut off from the Wall by the great hulk of the 1966 Staatsbibliothek, also by

Scharoun, its collections based on the part of the stock of the Prussian State Library that was evacuated westwards.

To list all West Berlin museums would be tedious; *Berlin Programm* for February 1987 listed no fewer than fifty, covering all conceivable subjects from the Police Museum to the Anti-war Museum. In the same month there were thirty-five special exhibitions, mostly in museums, not to speak of scores of exhibitions in the galleries of private dealers spread throughout the city. Two more museums out of many may be mentioned. The Berlin Museum, specializing in Berlin's history, occupies one of the few really old buildings in West Berlin, the restored Kammergericht, erected in 1735 for legal and administrative purposes. This collection on Berlin life and history even includes a reconstruction of a traditional Berlin *Kneipe* (tavern), which is so popular that it never seems possible to get in. Among other specializations, the museum has recently concentrated on building up a Jewish collection, replacing the Jewish Museum in the Oranienburger Strasse (in the present East Berlin) that was closed on 10 November 1938, the day after the neighbouring Great Synagogue had been set on fire. The new Transport and Technology Museum is an interesting attempt to recycle buildings on the southern approach to the former Anhalt Station.

A project for the future is a much discussed Museum of German History, probably to be located in the Martin-Gropius-Bau, a building originally erected for the Museum of Applied Arts, situated in the Wall Fringe adjacent to the site of the Gestapo headquarters in the Prinz-Albrecht-Strasse. The future treatment of this latter site, imbued with so many tragic memories, is also a matter of considerable controversy.

The extraordinarily active Berlin theatre of the 1920s, destroyed by National Socialism and war, has been followed by a post-war renaissance, although in the decentralized urban structure of present-day Germany, West Berlin's theatres have lost their automatic predominance; a production in Frankfurt or Bochum can attract national attention. Keystones of the theatrical structure are provided by two state theatres, both in Charlottenburg. The Deutsche Oper was founded in 1912 and had its great years in the 1920s, associated with such names as Richard Strauss, Erich Kleiber, or Otto Klemperer; it was rebuilt in 1961. The Schillertheater, originally founded as 'Theater der Stadt Charlottenburg' in 1907, was rebuilt in 1950–1 to become the West Berlin home of the classical repertoire. The Schlosspark-Theatre in Steglitz is also part of the state theatre system, financed on a scale of unbelievable generosity by British standards.

The greatest theatrical vitality is, however, said to be found in the 'private' (but often well subsidized) theatres. Maximum attention in the years since 1970 has been seized by Peter Stein's nonconformist

Schaubühne, in 1981 transferred from proletarian Kreuzberg to an expensive experimental theatre constructed within the shell of Erich Mendelssohn's 1928 Universum Cinema, among the bright lights of the Kurfürstendamm. The Theater des Westens in the heart of the central business district is the home of operetta and musicals. Satirical cabaret is very much part of Berlin life; in addition to the well established 'Stachelschweine' (porcupine) and 'Wühlmäuse' (vole) there is a swarm of others, often ephemeral, particularly in the 'alternative scene'. The Berlin senate gives financial support to film making in Berlin, which also since 1951 has had one of the leading international film festivals. The Deutsche Kinemathek, founded in 1962, has one of the largest film collections in the Federal Republic.

For those who take pleasure in the open air, West Berlin's parks, woods, and lakes are always available, if liable to be overcrowded on fine weekends; it is even possible to go skiing on the Teufelsberg in the Grunewald, with machine-made snow if need be. *Berlin Programm* lists fifty-one 'places of interest' to visit, including the Berlin Zoo and Aquarium. The 42-hectare Botanical Garden is of particular interest because a large part of it is arranged geographically; in a few metres it is possible to visit the Limestone Alps, the Carpathians, the Himalayas, an area of lowland bog, or a hay meadow, all with their characteristic flora. A surprising length of river, lake, and canal is available for pleasure-boat trips. West Berlin offers the usual day and half-day bus tours of the city, but beyond its boundaries only East Berlin and Potsdam are available on this basis. In recent years there has been a tendency to introduce more 'off-beat' tours, such as 'Berlin in the 1920s' or 'Berlin's industrialization'. The admirable free leaflets available from 'Berlin Information' also allow such tours to be done on a 'do-it-yourself' basis.

9.2.3 Alternative living

Until the building of the Berlin Wall in 1961, West Berlin was politically most concerned with external dangers. After that time it to some extent turned in on itself politically. Particularly in the years 1967–8, young people demonstrated and fought on the streets in favour of a more open university administration, against the alleged 'pseudo-democracy' of the Federal Republic, and against the war in Vietnam. Later 'student' protest centred on the urban renewal projects of the senate and included the occupation of apartment houses due for renovation. The forces of the young were strengthened by young men moving to West Berlin to escape military service.

Today the alternative scene is a less combative one. West Berlin more than any city of Germany is the place where there is a nonviolent search for alternative living. The principle is to opt out of the cash economy by the formation of collective or co-operative groups. There are collectives for craft and repair services, health-food shops, self-help, communication skills, residential management, medical services, legal services, ecological groups, actors, and many more. There are said to be in West Berlin 1,500 'alternative' projects, occupying between 10,000 and 15,000 (mostly young) people. It may be questioned whether such groups have a really independent existence, or whether they are parasitic on the economy of West Berlin as a whole.

9.3 The future of Berlin

West Berlin, it can be argued, so rigorously shut off by an unsympathetic neighbour, is no longer an essential component of German life. There are already those who describe it as 'provincial'; will it relapse into the condition of an 'old-age pensioner' of the Federal Republic, until the West Germans no longer care enough to keep it out of the hands of the GDR? Or will the initiative and creative ability inherent in a free economy find new and viable roles for the walled city?

Then there is East Berlin, which indisputably exercises capital functions in respect of the GDR. Here there is no question of decline; not only is power exercised in the way that power is not exercised in West Berlin, but the part-city is perceived by the inhabitants of the GDR as the place where things are happening, the goal of migratory movements by all the most ambitious and energetic elements in the population. Will the image of this expanding East Berlin eventually occlude the more familiar image of West Berlin, even in the perception of those who are not residents of the GDR?

Is life in isolation the only future scenario for West Berlin? There are those now who are prepared to think what was formerly unthinkable and envisage quite a different future for the city. In 1986 a few Social Democrats and supporters of the 'Alternative List' in the Berlin parliament put forward proposals for cutting West Berlin's political, economic, and financial links with the Federal Republic. The city would become a 'bridge of peace between East and West', no longer sundered from its 'socialist *umland*'.

Could West Berlin emerge as a free city, like inter-war Danzig (the analogy is scarcely encouraging)? Perhaps the fact that such a solution was a policy aim of Stalin and Krushchev, and firmly rejected by West Berlin's leaders, should not be a bar to reconsideration today? Even in the unlikely event of such a scheme being acceptable to majority opinion in West Berlin, it is difficult to see how, under present political and economic conditions, it could be acceptable without a degree of

GDR control over all movements, including air movements, to and from the city that would surely be unacceptable. Only a quite different international situation, with the emergence of a new 'Mitteleuropa' free of East–West conflicts, would seem to make possible a 'Free City of West Berlin'.

References

Items for which a fuller entry is given for an earlier chapter are marked as appropriate (ch. 1, ch. 2, etc.).

1 Berlin: product and victim of history (also items relating to more than one chapter)

Bader, F.J.W. and Müller, D.O. (eds) (1981) *Stadtgeographischer Führer Berlin (West)*, 2nd edn, Berlin and Stuttgart: Borntraeger (*Sammlung geographischer Führer* 7).

Cornish, V. (1923) *The Great Capitals: An Historical Geography*, London: Methuen.

Dickinson, R.E. (1945) *The Regions of Germany*, London: Kegan Paul, esp. ch. 9 (considerable historical interest).

—— (1953) *Germany: A General and Regional Geography*, London: Methuen, esp. ch. 26 (considerable historical interest).

Dietrich, R. (1960) *Berlin: neun Kapitel seiner Geschichte*, Berlin: de Gruyter.

Elkins, T.H. (1968) *Germany*, 2nd edn, London: Chatto & Windus; New York: Praeger, ch. 21.

—— (1969) 'Both sides of Berlin', *Geographical Magazine*, 41, pp. 382–92.

Engeli, C. (1986) *Landesplanung in Berlin-Brandenburg; eine Untersuchung zur Geschichte des Landesplanungsverbandes Brandenburg-Mitte 1929– 1936.* Stuttgart, etc.: Kohlhammer.

Hofmeister, B. (1975a) *Berlin: eine geographische Strukturanalyse der zwölf westlichen Bezirke (Bundesrepublik Deutschland und Berlin I)*, Darmstadt: Wissenschaftliche Buchgesellschaft.

—— (1975b) 'Berlin mit besonderer Berücksichtigung der Entwicklung von Berlin (West) in den Jahren 1961–73', *Geographisches Taschenbuch* (1975–76), Wiesbaden: Steiner.

—— (1978) 'Germany's industrial giant; isolated West Berlin', *Geographical Magazine*, 50, pp. 439–45.

Hofmeister, B., Pachur, H.-J., Pape, C., and Reindke, G. (eds) (1985) *Berlin: Beiträge zur Geographie eines Großstadtraumes*, Berlin: Reimer.

Hofmeister, B. and Voss, F. (eds) (1985) *Exkursionsführer zum 45. Deutschen Geographentag Berlin*, Berlin: Institut für Geographie der Technischen Universität Berlin (*Berliner geographische Studien* 17).

Im Überblick: Berlin (1987), 4th edn, Berlin: Informationszentrum Berlin. An English-language version of this publication is periodically updated and made available under various titles, including *In Brief: Berlin, Outlook Berlin,* and *750 Years Berlin 1970: Information.*

Kluczka, G. (1985) 'Berlin (West) – Grundlagen und Entwicklung', *Geographische Rundschau* 37: 422–36.

Leyden, F. (1933) *Gross-Berlin: Geographie der Weltstadt,* Breslau: Hirt.

Louis, H. (1936) 'Die geographische Gliederung von Gross Berlin', in H. Louis and W. Panzer (eds) *Länderkundliche Forschung: Festschrift zur Vollendung des 60. Lebensjahres Norbert Krebs,* Stuttgart: Engelhorn, 146–71.

Mann, G. (1968, 1974) *The History of Germany since 1789,* London: Chatto & Windus/Penguin.

Müller, H. (1985) 'Berlin (West) und Berlin (Ost); sozialräumliche Strukturen einer Stadt mit unterschiedlichen Gesellschaftssystemen', *Geographische Rundschau* 37: 437–41.

Pfannschmidt, M. (1971) 'Landesplanung Berlin-Brandenburg-Mitte', in *Raumordnung und Landesplanung im 20. Jahrhundert. Forschungs- und Sitzungsberichte der Akademie für Raumforschung und Landesplanung* 63: 29–54.

Schinz, A. (1964) *Berlin: Stadtschicksal und Städtebau,* Braunschweig: Westermann.

Schneider, W. (1980) *Berlin: eine Kulturgeschichte in Bildern und Dokumenten,* Leipzig and Weimar: Kiepenheuer.

Schulz, H. (ed.) (1987) *Sektion Geographie der Humboldt-Universität zu Berlin: Jubiläumskonferenz 1986* (Berliner Geographische Arbeiten, Sonderheft 4).

Vogel, W. (1966) *Führer durch die Geschichte Berlins,* Berlin: Rembrandt.

Werner, F. (1978) *Stadtplanung Berlin: Theorie und Realität,* 2nd edn, Berlin: Kiepert.

Specifically related to East Berlin

Kohl, H. *et al.* (1974, 1976) *Die Bezirke der Deutschen Demokratischen Republik,* Gotha and Leipzig: Haack (see ch. 'Berlin, die Hauptstadt der Deutschen Demokratischen Republik).

Rumpf, H. (1986) 'Die Hauptstadt Berlin und ihr Umland als Gegenstand geographischer Forschung', in Schulz (1986) (this chapter).

Schulz, H. (ed.) (1987) *Sektion Geographie der Humboldt-Universität zu Berlin: Jubiläumskonferenz 1986* (Berliner geographische Arbeiten, Sonderheft 4).

Zur Entwicklung Berlins 1970–1985. Berlin: Magistrat von Berlin, Hauptstadt der DDR, 1986.

2 Berlin divided

Anderhub, A., Bennett, J.O. and Reese, H.G. (1984) *Blockade, Airlift and Airlift Gratitude Foundation* (Berliner Forum, English edn 2/84), Berlin: Press and Information Office of the Land Berlin.

Berlin (Land) (1984) *Die Verfassung von Berlin und das Grundgesetz für die Bundesrepublik Deutschland, mit ergänzenden Dokumenten,* 14th edn,

Berlin: Landeszentrale für politische Bildungsarbeit.

Bucerius, G. (1985) 'Dann wären nur Scherben geblieben . . .', *Die Zeit*, 29 November.

Catudal, H.M. (1972) 'The Berlin agreements on exchange of exclaves and enclaves', *Geoforum* 11: 78–80.

—— (1978a) 'Problems and perspectives of the current Berlin settlement', *Geo Journal* 2: 5–9.

—— (1978b) *The Diplomacy of the Quadripartite Agreement on Berlin: A New Era in East–West Politics*, Berlin: Berlin Verlag.

—— (1978c) *A Balance Sheet of the Quadripartite Agreement on Berlin: Evaluation and Documentation*, Berlin: Berlin Verlag.

Childs, D. (1983) *The GDR: Moscow's German Ally*, London: Allen & Unwin.

Deighton, L. (1964) *Funeral in Berlin*, London: Cape (Penguin, 1966).

Engert, J. (1985) 'Berlin between east and west; lessons for a confused world' in Merritt and Merritt (1985) (this ch.).

Francisco, R.A. (1986) 'Introduction: divided Berlin in postwar politics', in Francisco and Merritt (1986) (this ch.).

Francisco, R.A. and Merritt, R.L. (ed.) (1985) *Berlin between Two Worlds*, Boulder, Colo., and London: Westview.

Heidelmeyer, W. (1970) *The Status of the Land Berlin*, Berlin: Press and Information Office of the Land Berlin.

Hillenbrand, M.J. (1981) *The Future of Berlin*, Monclair: Allenheld, Osmun.

Hofmeister, B. (1981) 'West Berlin and the Federal Republic of Germany', in G. W. Hoffman (ed.) *Federalism and Regional Development*, Austin, Tex.

Im Überblick: Berlin (1987) (ch. 1).

Leonhard, W. (1955) *Die Revolution entlässt ihre Kinder*, Köln: Kiepenheuer & Witsch. Translated in1957 by C. M. Woodhouse as *Child of the Revolution*, London: Collins.

Mander, J. (1962) *Berlin: Hostage for the West*, London: Penguin.

Merritt, R. L. (1973) 'Infrastructural changes in Berlin', *Annals of the Association of American Geographers* 63: 58–70.

—— (1985) 'Interpersonal transactions across the Wall', in Merritt and Merritt (1985) (this ch.).

—— (1986) 'Postwar Berlin: divided city', in Francisco and Merritt (1986) (this ch.).

Merritt, R. L. and Merritt, A. J. (eds) (1985) *Living with the Wall: West Berlin, 1961–1985*, Durham, NC: Duke University Press.

Mosely, P.E. (1950a) 'Dismemberment of Germany', *Foreign Affairs* 28: 487–98.

—— (1950b) 'The occupation of Germany', *Foreign Affairs* 28: 580–604.

Reindke, G. (1985) 'Die Milchversorgung von Berlin in Vergangenheit und Gegenwart' in Hofmeister *et al.* (1985) (ch. 1).

Schiedermair, H. (1975) *Der völkerrechtliche Status Berlins nach dem Viermächte-Abkommen vom 3. September 1971*, 1975, Berlin: Springer.

Schütz, K. (1985) 'The Wall and West Berlin's development', in Merritt and Merritt (1985) (this ch.).

Smith, J.E. (1963) *The Defense of Berlin*, Baltimore: Johns Hopkins Press; London: Oxford University Press.

Wettig, G. (1981) *Das Vier-Mächte-Abkommen in der Bewährungsprobe; Berlin im Spannungsfeld von Ost und West*, Berlin: Berlin Verlag.

—— (1986) 'The relations between West Berlin and the Warsaw Pact states', in Francisco and Merritt (1986) (this ch.).

256 | Berlin

Windsor, P. (1963) *City on Leave: A History of Berlin, 1945–62*, London: Chatto & Windus.

Specifically related to East Berlin

German Democratic Republic, Ministry of Foreign Affairs (1964a) *The Problem of West Berlin and Solutions proposed by the Government of the German Democratic Republic*, Berlin.
—— (1964b) *Documentation on the Question of West Berlin*, Dresden: Zeit im Bild.
Schöne, H. (1961) 'The status of Berlin' in Stulz (1961) (this ch.).
Stulz, P. (1961) 'The division of the German capital and the establishment of the democratic city council of Greater Berlin in November 1948', in Stulz (1961) (this ch.).
—— (ed.) (1961) *Berlin 1945–48: Contributions to the History of the City*, Leipzig: VEB Edition.

3 The Berlin countryside

Assmann, P. (1957) *Der geologische Aufbau der Gegend von Berlin*, Berlin: Senator für Bau- und Wohnungswesen.
Behrmann, W. (1949/50) 'Die Umgebung Berlins nach morphologischen Formengruppen betrachtet', *Die Erde* 1: 93–122.
Böse, M. (1979) *Die geomorphologische Entwicklung im westlichen Berlin nach neueren stratigraphischen Untersuchungen* (Berliner geographische Abhandlungen 28).
—— (1985) 'Aspekte zur Stadtmorphologie von Berlin (West)' in Hofmeister *et al.* (1985) (ch. 1).
Ergenzinger, P.J. (1985) 'Niederschläge von Staub und Schwermetallen in Berlin (West)', in Hofmeister *et al.* (1985) (ch. 1).
Fichtner, V. (1977) *Die anthropogen bedingte Umwandlung des Reliefs durch Trümmeraufschüttungen in Berlin (West) seit 1945* (Abhandlungen des 1. Geographischen Instituts der Freien Universität Berlin 21), Berlin: Selbstverlag des Geographischen Instituts der Freien Universität.
Haserodt, K. (1985) 'Stadtrand im Westen – das "andere" Berlin: Gatow–Kladow', in Hofmeister and Voss (1985) (ch. 1).
Kallenbach, H. (1980) 'Abriss der Geologie von Berlin', in *Beiträge zu den Tagungsunterlagen des Internationalen Alfred-Wegener-Symposiums und der Deutschen Meteorologen-Tagung 1980*, 15–21.
Karrasch, H. (1985) 'Luftqualität und Mortalität in Berlin (West)', in Hofmeister *et al.* (1985).
Leyden, F. (1933) See section 1.
Naturbuch Berlin: Pflanzen – Tiere – Lebensräume (1986) Berlin: Senator für Stadtentwicklung und Umweltschutz.
Pachur, H.J., Schulz, G., and Stäblein, G. (1985) 'Die geomorphologische Detailkartierung. Berichte über das GMK-Schwerpunktprogramm mit dem Beispielblatt GMK 25B-Zehlendorf 3545', in Hofmeister *et al.* (1985) (ch. 1).
Schlaak, P. (1980) 'Berliner meteorologische Daten aus drei Jahrhunderten und Ergebnisse von stadtweiten Messnetzen der vergangenen beiden

Jahrzehnte', in *Beiträge zu den Tagungsunterlagen des Internationalen Alfred-Wegener-Symposiums und der Deutschen Meteorologen-Tagung 1980*, 2–14.

Steinecke, A. (1985) 'Freizeit in der räumlich isolierten Grossstadt: Freiflächenausstattung und Freizeitverhalten in Berlin (West)', in Hofmeister *et al.* (1985) (ch. 1).

Sukopp, H. *et al.* (1980) *Beiträge zur Stadtökologie von Berlin (West): Exkursionsführer für das 2. Europäische ökologische Symposium*, Berlin: Technische Universität.

Treter, U. (1985) 'Grundwassernutzung und Grundwassererneuerung in Berlin (West)', in Hofmeister *et al.* (1985) (ch. 1).

Vollmar, R. (1985) 'Luftbilder von Lübars 1959 und 1979. Bildinterpretation und Nutzungskonflikte bei der Gestaltung von Dorf und Umland', in Hofmeister *et al.* (1985) (ch. 1).

Specifically related to East Berlin

Marcinek, J. and Nitz, B. (1973) *Das Tiefland der Deutschen Demokratischen Republik: Leitlinien seiner Oberflächengestaltung*, Gotha and Leipzig: Haack.

Marcinek, J., Saratka, J. and Zaumseil, L. (1983a) *Die natürlichen Verhältnisse der Hauptstadt der DDR, Berlin, und ihres Umlandes; ein Überblick*, Berlin: Magistrat von Berlin, Abteilung Volksbildung, Bezirkskabinett für Weiterbildung.

—— (1983b) 'Grundzüge der Flächennutzung und Umweltprobleme im Gebiet der Hauptstadt der DDR, Berlin', in Rumpf (1983) (this ch.).

Richter, H. and others (ed.) (1969) *Berlin: die Hauptstadt der DDR und ihr Umland*, Gotha and Leipzig: Haack.

Rumpf, H. (ed.) (1983) *Geographische Beiträge zur Entwicklung und Gestaltung territorialer Beziehungen zwischen der Hauptstadt der DDR, Berlin, und ihrem Umland*, Berlin: Humboldt-Universität zu Berlin, Sektion Geographie (*Berichte* 19/83).

Rumpf, H., Leupolt, B., Zaumseil, L., and Heerwagen, D. (1985) 'Entwicklungsaspekte der Stadt-Umland-Region der Hauptstadt Berlin unter besonderer Berücksichtigung des Umlandes', *Petermanns geographische Mitteilungen* 129: 111–19.

Sadler, W. (1986) 'Entwicklungsaspekte der regionalen Agrarstruktur in der Stadt-Umland-Region der Hauptstadt der DDR, Berlin', in Schulz (1986) (ch. 1).

Schulz, M. and Strehz, J.R. (1983) 'Zu einigen territorialen Aspekten der Naherholung der Berliner Bevölkerung in der Stadt-Umland-Region der Hauptstadt der DDR, Berlin', in Rumpf (1983) (this ch.).

Seibicke, W. (1983) 'Zu verkehrsgeographischen Grundlagen, Tendenzen und Problemen erholungsräumlicher Beziehungen zwischen der Hauptstadt der DDR, Berlin, und ihrem Umland unter besonderer Berücksichtigung des Wochenenderholungsverkehrs, in Rumpf (1983) (this ch.).

Spitzer, H. (1975) 'Zur Gestaltung von Naherholungsgebietskarten—Erläuterungen zur Übersichtskarte "Berlin und umliegendes Territorium"', *Wissenschaftliche Zeitschrift der Humboldt-Universität, Math.-Nat. Reihe* 24 (Berliner Geographische Arbeiten 60).

Zaumseil, L. (1986) 'Zur Kennzeichnung des Flächennutzungskomplexes Stadt-Umland-Region, dargestellt am Beispiel der Stadt-Umland-Region der Hauptstadt der DDR, Berlin', in Schulz (1986) (ch. 1).

Zaumseil, L., Marcinek, J., and Saratka, J. (1983) 'Eine Kurzdarstellung des Klimas von Berlin, Hauptstadt der DDR', in Rumpf (1983) (this ch.).

4 Transport developments

Beaver, S.H. (1937) 'The railways of great cities', *Geography* 22: 116–20. Reprinted in H.M. Mayer and C.F. Kohn (eds) *Readings in Urban Geography*, Chicago: Chicago University Press.

Elkins, T.H. (1984) 'S-Bahn goes west', *Geographical Magazine* 56: 166–8.

Hofmeister, B. (1975a) (ch. 1): 20–6, 225–66.

—— (1985) 'Alt-Berlin–Gross-Berlin–West Berlin: Versuch einer Flächennutzungsbilanz 1786–1985', in Hofmeister *et al.* (1985) (ch. 1).

Mayr, A. (1985) 'Berlin als Flughafenstandort', in Hofmeister *et al.* (1985) (ch. 1).

Wangemann, V. (1984) *Nahverkehrsentwicklung und Nahverkehrsplanung in Berlin (West) seit 1945*, Berlin: Reimer.

5 The Berlin economy

Bader, F.J.W. (1985a) 'Moabit–Industriestandort und Wohngebiet', in Hofmeister and Voss (1985) (ch. 1).

—— (1985b) 'Wohnen und Arbeiten auf dem Wedding', in Hofmeister and Voss (1985) (ch. 1).

'Ergebnisse der volkswirtschaftlichen Gesamtrechnungen für Berlin (West)' (1984), *Berliner Statistik* 39, (1985): 286–302.

Hofmeister, B. (1975a), pp. 168–225. See section 1.

—— (1980) 'Charlottenburg und die Entwicklung der City von West-Berlin' in W. Ribbe (ed.) *Von der Residenz zur City: 275 Jahre Charlottenburg*, Berlin: Colloquium Verlag.

—— (1981) 'Moabit: Durchgangsstation im Zuge der Randwanderung der Industrie?' in K. Schwarz (ed.)*Berlin: Von der Residenzstadt zur Industriemetropole 1* (Aufsätze), Berlin: Universitätsbibliothek der Technischen Universität.

Kluczka, G. (1985) See ch. 1.

Krafft, H. (1984) *Marktwirtschaft auf dem Prüfstand; 45 Jahre Berliner Wirtschaft*, Berlin and Offenbach: VDE-Verlag.

Müller, K.J. (1985) 'Die städtebauliche Entwicklung des Bezirks Wedding', in Hofmeister and Voss (1985) (ch. 1).

Reindke, G. (1985) 'Die Milchversorgung von Berlin in Vergangenheit und Gegenwart', in Hofmeister *et al.* (1985) (ch. 1).

Schütte, R. (1985) 'Berlin (West) im innerdeutschen Handel', *Berlin Information*, Januar.

Vetter, F. (1984) 'Recent changes in West Berlin's tourist flows', *Problem turystyki* 2: 40–6.

Watter, V. (1985) 'The West Berlin economy', in Merritt and Merritt (1985) (ch. 2).

Zimm, A. (1959) *Die Entwicklung des Industriestandortes Berlin*, Berlin: Deutscher Verlag der Wissenschaften.
—— (1961a) 'Westberlin – eine politisch-ökonomisch-geographische Dokumentation', *Einheit*, 9, September: 1354–72.
—— (1961b) *Westberlin. Der Industriestandort Westberlin unter den Bedingungen der Frontstadt: eine politisch- und ökonomisch-geographische Charakteristik*, Berlin: Deutscher Verlag der Wissenschaften.

Specifically related to East Berlin

Bräuniger, J. (1986) 'Development and organisation of tourism in Berlin, Capital of the GDR', in F. Vetter (ed.) *Grossstadttourismus*, Berlin: Reimer.
Flauss, H. (1977) 'Zur qualitativen und quantitativen Gestaltung der arbeitsräumlichen Stadt – Umland – Beziehungen am Beispiel der Hauptstadt der DDR, Berlin, *Wissenschaftliche Zeitschrift der Humboldt-Universität zu Berlin, Math.-Nat.*, Reihe 26 (6) (Berliner geographische Arbeiten 67: 747–54).
Kalisch, K.-H. (1980) 'Die Stadt-Umland-Region der Hauptstadt der DDR, Berlin; ein Beitrag zum Stadt–Umland–Problem', *Geographische Berichte* 25: 83–100.
Kehrer, G. (1986) 'Entwicklungstendenzen der Hauptstadt der DDR, Berlin, als Industrie- und Forschungszentrum', in Schulz (1986) (ch. 1).
Kehrer, G. and Fege, B. (1980) 'Zur Funktion und räumlichen Struktur der Hauptstadt der DDR, Berlin', *Zeitschrift für den Erdkundeunterricht* 32: 132–47.
Kohl, H. *et al.* (1974, 1976) (ch. 1).
Leupolt, B. (1983) 'Zur territorialen Industriestruktur des Umlandes der Hauptstadt der DDR, Berlin, und ausgewählte Aspekte produktionsräumliche Stadt-Umland-Beziehungen der Industrie', in Rumpf (1983) (this ch.).
—— (1986) 'Entwicklungstendenzen der Industriestruktur in der Stadt-Umland-Region der Hauptstadt Berlin', in Schulz (1986) (ch. 1).
Richter, H. (ed.) (1969) *'Berlin die Hauptstadt der DDR und ihr Umland*, (Wissenschaftliche Abhandlungen der Geographischen Gesellschaft in der DDR 10).
Rumpf, H. (ed.) (1983) *Geographische Beiträge zur Entwicklung und Gestaltung territorialer Beziehungen zwischen der Hauptstadt der DDR und ihrem Umland*, Berlin: Humboldt-Universität zu Berlin, Sektion Geographie (*Berichte* 19 (1983)).
Rumpf, H., Leupolt, B., Zaumseil, L., and Heerwagen, D. (1985) 'Entwicklungsaspekte der Stadt–Umland–Region der Hauptstadt Berlin unter besonderer Berücksichtigung des Umlandes', *Petermanns geographische Mitteilungen* 129: 111–19.
Sadler, W. (1983) 'Zur territorialen Agrarstruktur des Umlandes der Hauptstadt der DDR, Berlin, und ausgewählte Aspekte produktionsräumlicher Stadt-Umland-Beziehungen der Landwirtschaft', in Rumpf (1983) (this ch.).
—— (1986) 'Entwicklungsaspekte der regionalen Agrarstruktur in der Stadt-Umland-Region der Hauptstadt der DDR, Berlin', in Schulz (1986) (ch. 1).
Zimm, A. (1981) 'Zur Dynamik der Territorialstruktur der DDR-Hauptstadt

Berlin und ihres Umlandes', *Petermanns geographische Mitteilungen* 125: 157–65.

—— (1982) 'Zur strukturellen und funktionellen Dynamik in den typologischen Subräumen des Wirtschafts- und Lebensgebietes der Hauptstadt der DDR, Berlin', *Sitzungsberichte der Akademie der Wissenschaften der DDR, Mathematik – Naturwissenschaft – Technik*, 9N.

—— (1965) 'Vergleichende Funktionsanalyse des Demokratischen Berlins und Westberlins', *Petermanns geographische Mitteilungen* 109: 194–207.

Zimm, A. and Bräuniger, J. (1984) 'The agglomeration of the GDR capital, Berlin: a survey of its economic geography', *GeoJournal* 8: 23–31.

6 Urban development

Bader, F.J.W. (1985c) 'Der Wilhelminische Ring in Berlin und seine Entwicklung', in Hofmeister *et al.* (1985) (ch. 1).

Berlin (West): Senator für Bau- und Wohnungswesen (1981) *Preussische Bauten in Berlin*.

Boberg, J. *et al.* (eds) (1984) *Exerzierfeld der Moderne; Industriekultur in Berlin im 19. Jahrhundert, v. 1*, Munich: Beck.

Burgess, E.W. (1925) 'The growth of the city: an introduction to a research project', in R.E. Park, E.W. Burgess, and R.D. McKenzie, R.D. (eds) *The City*, Chicago: University of Chicago Press. Reprinted (1961) in G.A. Theodorson (ed.) *Studies in Human Ecology*, Evanston and New York: Harper & Row.

Conzen, M.R.G. (1960) *Alnwick, Northumberland: A Study in Town-Plan Analysis* (Transactions Institute of British Geographers 27).

Engeli, Ch. (1986) See ch. 1.

Escher, F. (1985) *Berlin und sein Umland; zur Genese der Berliner Stadtlandschaft bis zum Beginn des 20. Jahrhunderts*, Berlin: Colloquium Verlag (Einzelveröffentlichungen der Historischen Kommission zu Berlin 47).

Harris, C.D. and Ullman, E.L. (1945) 'The nature of cities', *Annals of the American Academy of Political and Social Science* 242: 7–17. Reprinted in H.M. Mayer and C.F. Kohn (ed.) *Readings in Urban Geography*, Chicago: University Press.

Heinrich, E., Mielke, F. *et al.* (1964) *Berlin und seine Bauten II; Rechtsgrundlagen und Stadtentwicklung*, Berlin and Munich: Ernst.

Hofmann, W. (1978) 'Wachsen Berlins im Industriezeitalter; Siedlungsstruktur und Verwaltungsgrenzen', in H. Jäger (ed.) *Probleme des Städtewesens im industriellen Zeitalter* (= Städteforschung Reihe A, Darstellungen Bd. 5), Cologne and Vienna: Böhlau).

Hofmeister, B. (1961) 'Das Problem der Nebencities in Berlin', *Berichte zur deutschen Landeskunde* 28: 45–69.

—— (1975a) See ch. 1.

—— (1980) See. ch. 5.

—— (1983) 'Die Siedlungsentwicklung Groß-Berlins' in K. Fehn (ed.) *Siedlungsforschung: Archäologie – Geschichte – Geographie* 1: 39–63.

—— (1985) 'Alt-Berlin – Groß-Berlin – West Berlin: Versuch einer Flächennutzungsbilanz 1786–1985', in Hofmeister *et al.* (1985) (ch. 1).

—— (1986) 'Wilhelminischer Ring und Villenkoloniengründung. Sozioökonomische und planerische Hintergründe simultaner städtebaulicher Prozesse im Grossraum Berlin 1860–1920', in H. Heineberg (ed.) *Innerstädtische Prozesse im 19. und 20. Jahrhundert: geographische und historische Aspekte* (Reihe Städteforschung).

Hoyt, H. (1939) 'The pattern of movement of residential rental neighbourhoods', in *The Structure and Growth of Residential Neighbourhoods in American Cities*, Washington DC: Federal Housing Administration. Reprinted in H.M. Mayer and C.F. Kohn (ed.) *Readings in Urban Geography*, Chicago: Chicago University Press.

Huse, N. (ed.) (1985) *Siedlungen der zwanziger Jahre – heute; vier Berliner Großsiedlungen 1924–84*, Berlin: Bauhaus-Archiv.

Junghanns, K. (1970) *Bruno Taut 1880–1938*, Berlin: Elefanten.

Kloss, K.-P. (1982) *Siedlungen der 20er Jahre*, Berlin: Haude and Spener.

Leyden, F. (1933) See ch. 1.

Louis, H. (1936) See ch. 1.

Matzerath, H. (1981) 'The influence of industrialization on urban growth in Prussia (1815–1914)', in H. Schmal (ed.) *Patterns of European Urbanization since 1500*, London: Croom Helm.

—— (1984) 'Berlin 1890–1940', in A. Sutcliffe (ed.) *Metropolis 1890–1940* (Studies in History, Planning, and the Environment), London: Mansell.

Müller, D.O. (1978) *Verkehrs- und Wohnstrukturen in Gross-Berlin 1880–1980: geographische Untersuchungen ausgewählter Schlüsselgebiete beiderseits der Ringbahn* (Berliner geographische Studien 4).

Müller, K.J. (1985b) 'Zersiedlung und Ortsbildveränderung in der "Gartenstadt" Berlin-Frohnau', in Hofmeister *et al.* (1985) (ch. 1).

Mullin, J.R. (1982) 'Ideology, planning theory and the German city in the inter-war years', *Town Planning Review* 53: 115–30, 257–72.

Pape, Ch. (1985) 'Köpenick; Entwicklungsskizze eines Berliner *Bezirks*', in Hofmeister *et al.* (1985) (ch. 1).

Pitz, H., Hofmann, W., and Tomisch, J. (1984) *Berlin-W; Geschichte und Schicksal einer Stadtmitte. Vol. 1: Von der Preussischen Residenz zur geteilten Metropole. Vol. 2: Vom Kreuzberg-Denkmal zu den Zelten*, Berlin: Siedler. (The second volume gives a detailed documentary treatment of three key areas of inner Berlin.)

Posener, J. (1979) *Berlin auf dem Wege zu einer neuene Architektur: das Zeitalter Wilhelms II*, Munich: Prestel.

Rave, R. and Knöfel, H.-J. (1981) *Bauen zeit 1900 in Berlin*, 3rd edn, Berlin: Kiepert.

Reissner, A. (1984) *Berlin 1675–1945: The Rise and Fall of a Metropolis*, London: Wolff.

Ribbe, W. (ed.) (1980) *Von der Residenz zur City: 275 Jahre Charlottenburg*, Berlin: Colloquium.

Schinz, A. (1964) See ch. 1.

Schulz, J. and Gräbner, W. (1981) *Architekturführer DDR: Berlin, Hauptstadt der Deutschen Demokratischen Republik*, 3rd edn, Berlin: VEB Verlag für Bauwesen.

Thienel, I. (1973) *Städtewachstum im Industrialisierungsprozess des 19. Jahrhunderts; – das Berliner Beispiel*, Berlin: de Gruyter (Veröffentlichungen der Historischen Kommission zu Berlin 39).

Thienel-Saage, I. (1983) *Städtewachstum in der Gründerzeit: Beispiel Berlin*, Paderborn: Schöningh.

Vogel, W. (1966) *Führer durch die Geschichte Berlins*, Berlin: Rembrandt.

Voll, D. (1985a) 'Die Bedeutung privater Investitionen im Bereich der Verkehrsinfrastruktur durch Siemens in den Jahren 1899–1930 in Berlin-Siemensstadt', in Hofmeister *et al.* (1985) (ch. 1).

Werner, F. (1976) *Stadtplanung Berlin; Theorie und Realität. Pt 1: 1900–60*, Berlin: Kiepert.

Wolters, R. (1978) *Stadtmitte Berlin: Stadtbauliche Entwicklungsphasen von den Anfängen bis zur Gegenwart*, Tübingen: Wasmuth.

Wuthe, K. (1984) 'Berlin: von der Doppelstadt zur geteilten Stadt, eine historisch-geographische Betrachtung', *Geographische Berichte* 37 (1985): 422–7.

7 Urban development and redevelopment after 1945

Aust, B. (1970) *Stadtgeographie ausgewählter Sekundärzentren in Berlin (West)*, Berlin: Reimer (Abhandlungen des 1. Geographischen Instituts der Freien Universität Berlin 16).

—— (1972) 'Anwendungsmöglichkeiten stadtgeographischer Untersuchungen über Sekundärzentren in Berlin (West)', *Die Erde* 103: 295–301.

—— (1981) 'Das Einkaufszentrum Steglitz – Schlossstrasse', in F.J.W. Bader and D.O. Müller (1981) (ch. 1).

Bader, F.J.W. (1985a) 'Moabit: Industriestandort und Wohngebiet', in Hofmeister and Voss (1985) (ch. 1).

—— (1985b) 'Wohnen und Arbeiten auf dem Wedding', in Hofmeister and Voss (1985) (ch. 1).

Butzin, B. and Heineberg, H. (1980) 'Nutzungswandel und Entwicklungsprobleme integrierter Shopping-Center in West Berlin', in H. Heineberg (ed.) *Einkaufszentren in Deutschland; Entwicklung, Forschungsstand und Probleme*, Paderborn: Schöningh (Münsterische geographische Arbeiten 5).

Heineberg, H. (1977) *Zentren in West- und Ost-Berlin. Untersuchungen zum Problem der Erfassung und Bewertung großstädtischer funktionaler Zentrenausstattung in beiden Wirtschafts- und Gesellschaftssytemen Deutschlands*, Paderborn: Schöningh (Bochumer geographische Arbeiten, Sonderreihe 9).

—— (1979a) 'Service centres in East and West Berlin', in R.A. French and F.E.I. Hamilton (eds) *The Socialist City: Spatial Structure and Urban Policy*, Chichester: Wiley.

—— (1979b) 'West–Ost-Vergleich grossstädtischer Zentrenausstattungen am Beispiel Berlins', *Geographische Rundschau* 31: 434–43.

—— (1985a) 'Jüngere Wandlungen in der Zentrenausstattung Berlins im West–Ost Vergleich', in Hofmeister *et al.* (1985) (ch. 1).

—— (1985b) 'Zentren in Berlin (West) und Berlin (Ost); die beiden Stadtzentren und ausgewählte Nebengeschäftszentren im Vergleich', in Hofmeister and Voss (1985) (ch. 1).

—— (1986) 'Aspects of city-centre development in the "Two Germanies": the examples of West and East Berlin', in W. Ritchie, J.C. Stone, and A.S. Mather (eds), *Essays for Professor R.E.H. Mellor*, Aberdeen: University of Aberdeen.

Pfannschmidt, M. (1959) 'Probleme der Weltstadt Berlin', in J.H. Schultze (ed.) *Zum Problem der Weltstadt*, Berlin: de Gruyter.

Pirch, M. (1985) 'Strategien zur Stadterhaltung und -erneuerung am Beispiel des Sanierungsgebietes Kreuzberg-Kottbusser Tor', in Hofmeister and Voss (1985) (ch. 1).

Schöller, P. (1953) 'Stadtgeographische Probleme des geteilten Berlin', *Erdkunde* 7: 1–11.

—— (1974) 'Paradigma Berlin', *Geographische Rundschau* 26: 425–34.

Voll, D. (1985b) 'Das Märkische Viertel: Entstehung und Struktur einer Grosssiedlung in Berlin (West)', in Hofmeister and Voss (1985) (ch. 1).

Wiek, K. (1967) *Kurfürstendamm und Champs-Elysées; geographischer Vergleich zweier Weltstrassengebiete* (Abhandlungen des 1 Geographischen Instituts der Freien Universität Berlin 11).

Specifically related to East Berlin

Bartmann-Kompa, I., Kutschmar, A., Karn, H. *et al.* (1981) *Architekturführer DDR: Bezirk Potsdam*, Berlin: VEB Verlag für Bauwesen.

Flauss, H. (1977) 'Zur qualitativen und quantitativen Gestaltung der arbeitsräumlichen Stadt – Umland Beziehungen am Beispiel der Hauptstadt der DDR, Berlin', *Wissenschaftliche Zeitschrift der Humboldt-Universität zu Berlin, Math.-nat.* Reihe 26 (Berliner geographische Arbeiten 67): 747–54.

Heerwagen, D. (1986) 'Entwicklungstendenzen und -probleme ländlicher Teilräume der Stadt-Umland-Region der Hauptstadt der DDR, Berlin', in Schulz (1986) (ch. 1).

Kalisch, K.-H. (1980) 'Die Stadt – Umland – Region der Hauptstadt der DDR, Berlin: ein Beitrag zum Stadt – Umland Problem', *Geographische Berichte* 25: 83–100.

Korn, R. (1986) 'Das Bild unserer Hauptstadt verändert sich', *Architektur der DDR* 35: 201–11.

Lammert, U., Kadatz, H.J., Collein, E., and Gericke, H. (1969) *Architektur und Städtebau in der DDR*, Leipzig: Seeman Verlag für Deutsche Bauakademie Berlin.

Lüdemann, H., Grimm, F., Krönert, R., and Neumann, H. (eds) (1979) *Stadt und Umland in der Deutschen Demokratischen Republik*, Gotha and Leipzig: Haack (*Petermanns geographische Mitteilungen*, Ergänzungs-Heft 279).

Schöller, P. (1986) *Städtepolitik, Stadtumbau und Stadterhaltung in der DDR*, Wiesbaden: Steiner (Erdkundliches Wissen 81).

Schulz, J. and Gräbner, W. (1981) *Architekturführer DDR*, Berlin (ch. 6).

Spitzer, H. (1985) 'Arbeitspendlerbeziehungen im Bereich der Hauptstadt der DDR Berlin und ihres Umlandes: Erläuterung zur Arbeitspendlerkarte', *Zeitschrift für den Erdkundeunterricht*, 37, 17–23.

Wolf, W. (1986) 'Haupttendenzen der Entwicklung der sozialen Infrastruktur (Grundausstattung) im Umland der Hauptstadt der DDR, Berlin'. In Schulz (1986) (ch. 1).

Zache, M. and Freyer, B. (1986) 'Grundlinien zur städtebaulich-architektonischen Gestaltung der Hauptstadt der DDR, Berlin', *Architektur der DDR* 35: 219–27.

Zimm, A. and Bräuniger, J. (1984) See ch. 5.

8 The Berlin population

'Ausländer in Berlin (West) 1982 gegenüber 1973' (1984) *Berliner Statistik* 38: 228–44.

Backé, B. (1984) 'Bevölkerungsprobleme von Berlin (West) und ihre Bedeutung für die Stadtentwicklung', *Berichte zur deutschen Landeskunde* 58: 325–56.

'Bevölkerungsentwicklung in Berlin (West) 1984' (1985) *Berliner Statistik* 39: 264–71.

Müller, D.O. (1985) 'Zur Bevölkerungsentwicklung im Ballungsgebiet Berlin seit 1939', in Hofmeister *et al.* (1985) (ch. 1).

Müller, H. (1985) See ch. 1.

Scholz, F. (1985) 'Räumliche Ausbreitung türkischer Wirtschaftsaktivitäten in Berlin. Ein Beitrag zur Integrationsfrage der Türken', in Hofmeister *et al.* (1985) (ch. 1).

Vernon, J. (1984) 'Essor et declin d'une ville', *Population* 38: 866–71.

Specifically related to East Berlin

Flauss, H. and Schulz, M. (1983) 'Zur Entwicklung der Bevölkerungsstruktur der Hauptstadt der DDR, Berlin, unter besonderer Berücksichtigung der östlichen Stadtbezirke', in Rumpf (1983) (ch. 5).

Schulz, M. (1986) 'Zu Wirkungen und Entwicklungstendenzen des Migrationsprozesses in der Stadt-Umland-Region der Hauptstadt der DDR, Berlin', in Schulz (1985) (ch. 1).

Additional sources of information

Maps and atlases

Atlas von Berlin (Deutscher Planungatlas), ed. W. Behrmann, Berlin: Senator für Bau und Wohnungswesen.

Berlin im Kartenbild; zur Entwicklung der Stadt 1650–1950, ed. K. Lindner, Berlin: Staatsbibliothek Preussischer Kulturbesitz (Austellungskataloge 15), 1981.

Computeratlas Berlin, by G. Braun *et al.*, Berlin: Reimer, 1983.

Generalkarte Berlin und Umgebung 1:200,000, Stuttgart: Mairs Geographischer Verlag.

Historischer Handatlas von Brandenburg und Berlin, Berlin: de Gruyter, 1962–80.

Seydlitz Weltatlas, Berlin: CVK and Schroedel, 1984, pp. 38–9.

Topographischer Atlas Berlin, ed. U. Freitag and Ch. Pape, Berlin: Reimer, 1987, for Senator für Bau- und Wohnungswesen (Abteilung Vermessung) Berlin.

Übersichtskarte von Berlin (West) 1:50,000, Berlin: Senator für Bau und Wohnungswesen, 1980.

Übersicht über die Flächennutzungsplanung 1:50,000, 4th edn, Berlin: Senator für Bau und Wohnungswesen, 1983.

Umweltatlas Berlin, Berlin: Senator für Stadtentwicklung und Umweltschutz, 1985.

Wahlatlas Berlin 1981, Berlin: Geographisches Institut der Freien Universität (Manuskripte vol. 6), 1981.

Statistical sources

Statistisches Jahrbuch Berlin (annual) Berlin: Statistisches Landesamt (contains a section on East Berlin).

Berliner Statistik (monthly) Berlin: Statistisches Landesamt (each monthly part contains updated statistics and articles on a variety of subjects).

Statistisches Jahrbuch der Deutschen Demokratischen Republik (annual) Berlin: Staatsverlag der Deutschen Demokratischen Republik, for Staatliche Zentralverwaltung für Statistik.

Bibliographies of Berlin

Berlin-Bibliographie (bis 1960), ed. H. Zopf and G. Heinrich, Berlin: de Gruyter, 1965 (*Veröffentlichungen der Historischen Kommission zu Berlin* 15, 1965).

(1961–6), ed. U. Scholz and R. Stromeyer, Berlin: de Gruyter, 1973 (*Veröffentlichungen der Historischen Kommission zu Berlin* 43, 1973).

(1967–77), ed. U. Scholz and R. Stromeyer, Berlin: de Gruyter, 1984 (*Veröffentlichungen der Historischen Kommission zu Berlin* 58, 1984).

Berlin in Geschichte und Gegenwart; Jahrbuch des Landesarchivs Berlin (annual), ed. H.J. Reichart, Berlin: Medusa, 1983; thereafter Siedler (contains selected bibliography for the previous year, a chronicle of events (generally also one for a previous year in the period 1961–81), surveys of cultural activities and articles on Berlin history).

Berlin – Hauptstadt der DDR in Buch und Zeitschrift. Nebst Beilage: Literatur über Westberlin (annual), Berlin: Stadtbibliothek und Fachabteilung Ratsbibliothek.

Bibliographie zur Geschichte der Mark Brandenburg (annual, 1970–   ), ed. H. J. Schreckenbach (*Veröffentlichungen des Staatsarchivs Potsdam*).

Bibliographie zur Kunstgeschichte von Berlin und Potsdam, ed. S. Badstüber-Gröger, Berlin: Akademie Verlag, 1968.

Items relating to Berlin are also to be found in a number of general bibliographies, for example:

Deutschland-Archiv: Zeitschrift für Fragen der DDR und der Deutschlandpolitik (monthly) (as well as articles and reviews it carries twice a year a report on the principal items published in the Federal Republic and the GDR on Berlin and the GDR).

Sperling, W. (1978) *Landeskunde DDR; eine kommentierte Auswahlbibliographie*, Munich and New York: Saur. (*Bibliographien zur Regionalen Geographie und Landeskunde*, ed. W. Sperling and L. Zögner 1). Also *Ergänzungsband 1978–1983* (*Bibliographien zur Regionalen Geographie und Landeskunde* 5, 1984).

Neues Schrifttum zur deutschen Landeskunde (annual, 1978–   ) (Teilband 1: Zeitschriften; Teilband 2: Monographien, Hochschulschriften, Sammelschriften, Beiträge aus Sammelschriften, Nachträge), (from 1982) Teilband 3: Orte, ed. W. Sperling, Trier: Zentralausschuss für deutsche Landeskunde.

Index

Page references in *italic type* relate to figures, photographs or tables. (W) indicates that the reference relates to West Berlin, (E) to East Berlin, and (F) to the agglomeration field in the GDR outside the boundary of Greater Berlin.